# STUDENT WORKBOOK
## VOLUME 1

THIRD EDITION

# college physics
## a strategic approach

knight • jones • field

RANDALL D. KNIGHT
CALIFORNIA POLYTECHNIC STATE UNIVERSITY, SAN LUIS OBISPO

PEARSON

Boston Columbus Indianapolis New York San Francisco Upper Saddle River
Amsterdam Cape Town Dubai London Madrid Milan Munich Paris Montréal Toronto
Delhi Mexico City São Paulo Sydney Hong Kong Seoul Singapore Taipei Tokyo

Publisher: Jim Smith
Executive Editor: Becky Ruden
Vice-President of Marketing: Christy Lesko
Managing Development Editor: Cathy Murphy
Senior Development Editor: Alice Houston
Program Manager: Katie Conley
Project Manager: Martha Steele
Senior Administrative Assistant: Cathy Glenn
Full-Service Production and Composition: PreMediaGlobal
Manufacturing Buyer: Jeff Sargent
Marketing Manager: Will Moore
Senior Market Development Manager: Michelle Cadden
Market Development Coordinator: Kait Nagi
Cover Designer: Tandem Creative, Inc.
Cover Photo Credit: Borut Trdina/Getty Images
Printer: RR Donnelley

ISBN 10: 0-321-90886-4; ISBN 13: 978-0-321-90886-5

11 16

www.pearsonhighered.com

# Table of Contents

# Preface

It is highly unlikely that one could learn to play the piano by only reading about it. Similarly, reading physics from a textbook is not the same as doing physics. To develop your ability to do physics, your instructor will assign problems to be solved both for homework and on tests. Unfortunately, it is our experience that jumping right into problem solving after reading and hearing about physics often leads to poor "playing" techniques and an inability to solve problems for which the student has not already been shown the solution (which isn't really "solving" a problem, is it?). Because improving your ability to solve physics problems is one of the major goals of your course, time spent developing techniques that will help you do this is well spent.

Learning physics, as in learning any skill, requires regular practice of the basic techniques. That is what this *Student Workbook* is all about. The workbook consists of exercises that give you an opportunity to practice techniques and strengthen your understanding of concepts presented in the textbook and in class. These exercises are intended to be done on a daily basis, right after the topics have been discussed in class and are still fresh in your mind. Successful completion of the workbook exercises will prepare you to tackle the more quantitative end-of-chapter homework problems in the textbook.

You will find that many of the exercises are *qualitative* rather than *quantitative.* They ask you to draw pictures, interpret graphs, use ratios, write short explanations, or provide other answers that do not involve calculations. A few math-skills exercises will ask you to explore the mathematical relationships and symbols used to quantify physics concepts but do not require a calculator. The purpose of all of these exercises is to help you develop the basic thinking tools you'll later need for quantitative problem solving. It is highly recommended that you do these exercises *before* starting the end-of-chapter problems.

One example from Chapter 4 illustrates the purpose of this *Student Workbook.* In that chapter, you will read about a technique called a "free-body diagram" that is helpful for solving problems involving forces. Sometimes, students mistakenly think that the diagrams are used by the instructor only for teaching purposes and may be abandoned once Newton's laws are fully understood. On the contrary, professional physicists with decades of problem-solving experience still routinely use these diagrams to clarify the problem and set up the solution. Many of the other techniques practiced in the workbook, such as ray diagrams, graphing relationships, sketching field lines and equipotentials, etc., fall in the same category. They are used at all levels of physics, not just as a beginning exercise. And many of these techniques, such as analyzing graphs and exploring multiple representations of a situation, have important uses outside of physics. Time spent practicing these techniques will serve you well in other endeavors.

You will find that the exercises in this workbook are keyed to specific sections of the textbook in order to let you practice the new ideas introduced in that section. You should keep the text beside you as you work and refer to it often. You will usually find Tactics Boxes, figures, or examples in the textbook that are directly relevant to the exercises. When asked to draw figures or diagrams, you should attempt to draw them so that they look much like the figures and diagrams in the textbook.

Because the exercises go with specific sections in the text, you should answer them on the basis of information presented in *just* that section (and prior sections). You may have learned new ideas in Section 7 of a chapter, but you should not use those ideas when answering questions from Section 4. There will be ample opportunity in the Section 7 exercises to use that information there.

You will need a few "tools" to complete the exercises. Many of the exercises will ask you to *color code* your answers by drawing some items in black, others in red, and perhaps yet others in blue. You need to purchase a few colored pencils to do this. The authors highly recommend that you work in pencil, rather than ink, so that you can easily erase. Few are the individuals who make so few mistakes as to be able to work in ink! In addition, you'll find that a small, easily carried six-inch ruler will come in handy for drawings and graphs.

As you work your way through the textbook and this workbook, you will find that physics is a way of *thinking* about how the world works and why things happen as they do. We will primarily be interested in finding relationships, seeking explanations, and developing techniques to make use of these relationships, only secondarily in computing numerical answers. In many ways, the thinking tools developed in this workbook are what the course is all about. If you take the time to do these exercises regularly and to review the answers, in whatever form your instructor provides them, you will be well on your way to success in physics.

**To the instructor:** The exercises in this workbook can be used in many ways. You can have students work on some of the exercises in class as part of an active-learning strategy. Or you can do the same in recitation sections or laboratories. This approach allows you to discuss the answers immediately, to answer student questions, and to improvise follow-up exercises when needed. Having the students work in small groups (two to four students) is highly recommended.

Alternatively, the exercises can be assigned as homework. The pages are perforated for easy tear-out, and the page breaks are in logical places so that you can assign the sections of a chapter that you would likely cover in one day of class. Exercises should be assigned immediately after presenting the relevant information in class and should be due at the beginning of the next class. Collecting them at the beginning of class, and then going over two or three that are likely to cause difficulty, is an effective means of quickly reviewing major concepts from the previous class and launching a new discussion.

If used as homework, it is *essential* for students to receive *prompt* feedback. Ideally, this would occur by having the exercises graded, with written comments, and returned at the next class meeting. Posting fairly detailed answers on a course website also works. Lack of prompt feedback can negate much of the value of these exercises. Placing similar qualitative/graphical questions on quizzes and exams, and telling students at the beginning of the term that you will do so, encourages students to take the exercises seriously and to check the answers.

Student feedback from end-of-term questionnaires reveals three prevalent attitudes toward the workbook exercises:

i. They think it is an unreasonable amount of work.
ii. They agree that the assignments force them to keep up and not get behind.
iii. They recognize, by the end of the term, that the workbook is a valuable learning tool.

However you choose to use these exercises, they will significantly strengthen your students' conceptual understanding of physics.

Following the workbook exercises are optional Dynamics Worksheets, Momentum Worksheets, and Energy Worksheets for use with end-of-chapter problems in Parts I and II of the textbook. Their use is recommended to help students acquire good problem-solving habits early in the course. If you wish your students to use these, have them make enough photocopies for use throughout the term.

Answers to all workbook exercises are provided as pdf files on the Instructor Resource DVD and can also be downloaded from the Instructor Resource Area in MasteringPhysics®.

**Acknowledgments:** The author would like to thank Scott Nutter for his assistance in writing the solutions and Jared Sterzer and his colleagues at PreMediaGlobal for their production of the workbook.

# 1 Representing Motion

## 1.1 Motion: A First Look

**Exercises 1–5:** Draw a motion diagram for each motion described below.

- Use the particle model to represent the object as a particle.
- Six to eight dots are appropriate for most motion diagrams.
- Number the positions in order, as shown in Figure 1.4 in the text.

1. A car accelerates forward from a stop sign. It eventually reaches a steady speed of 45 mph.

2. An elevator starts from rest at the 100th floor of the Empire State Building and descends, with no stops, until coming to rest on the ground floor. (Draw this one *vertically* because the motion is vertical.)

3. A skier starts *from rest* at the top of a 30° snow-covered slope and steadily speeds up as she skies to the bottom. (Orient your diagram as seen from the *side*. Label the 30° angle.)

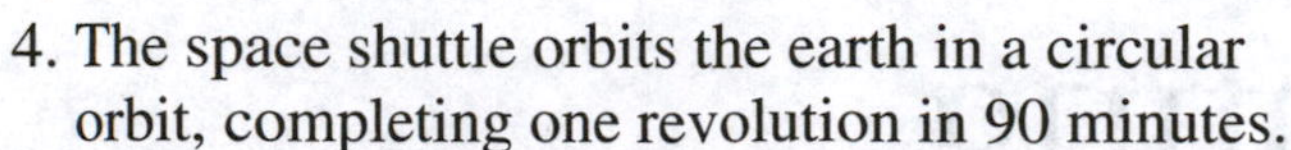

4. The space shuttle orbits the earth in a circular orbit, completing one revolution in 90 minutes.

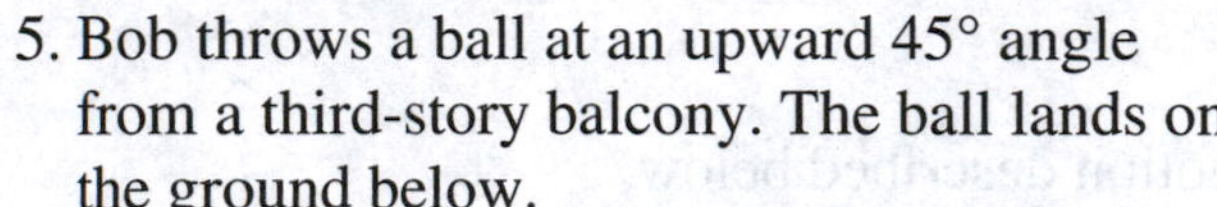

5. Bob throws a ball at an upward 45° angle from a third-story balcony. The ball lands on the ground below.

**Exercises 6–9:** For each motion diagram, write a short description of the motion of an object that will match the diagram. Your descriptions should name *specific* objects and be phrased similarly to the descriptions of Exercises 1 to 5. Note the axis labels on Exercises 8 and 9.

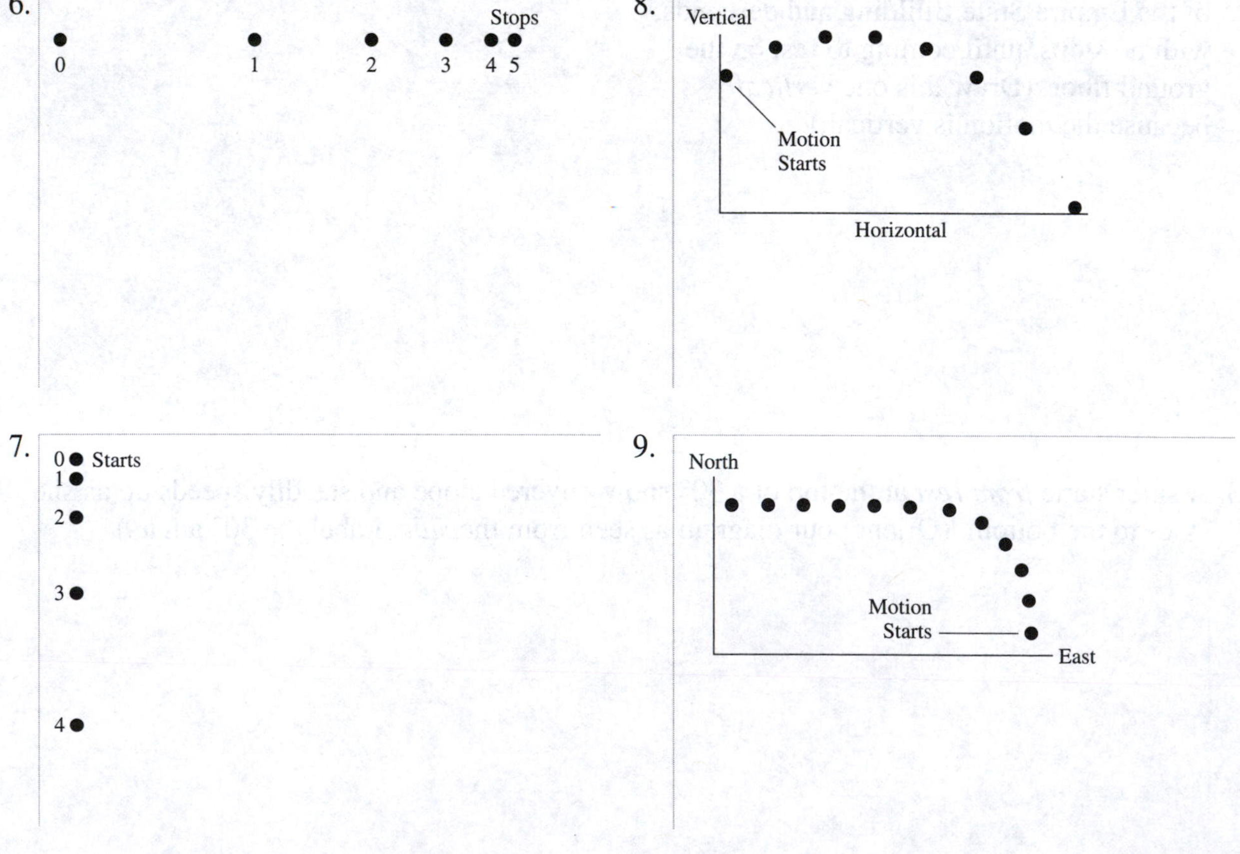

## 1.2 Position and Time: Putting Numbers on Nature

10. Redraw each of the motion diagrams from Exercises 1 to 3 in the space below. Add a coordinate axis to each drawing and label the initial and final positions. Draw an arrow on your diagram to represent the displacement from the beginning to the end of the motion.

11. In the picture below, Joe starts walking slowly but at constant speed from his house on Main Street to the bus stop 200 m down the street. When he is halfway there, he sees the bus and steadily speeds up until he reaches the bus stop.
    a. Draw a motion diagram in the street of the picture to represent Joe's motion.
    b. Add a coordinate axis below the picture with Joe's house as the origin. Label Joe's initial position at the start of his walk as $x_1$, his position when he sees the bus as $x_2$, and his final position when he arrives at the bus stop as $x_3$. Draw arrows above the motion diagram to represent Joe's displacement from his initial position to his position when he first sees the bus and the displacement from where he sees the bus to the bus stop. Label these displacements $\Delta x_1$ and $\Delta x_2$, respectively.

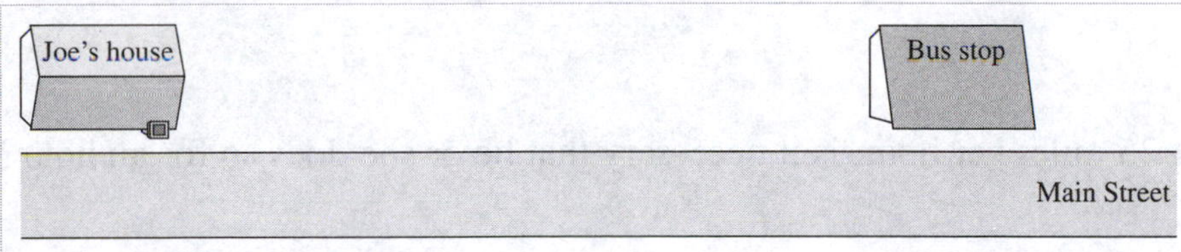


    c. Repeat part b in the space below but with the origin of the coordinate axis at the location where Joe starts to speed up.

    d. Do the displacement arrows change when you change the location of the origin?

## 1.3 Velocity

12. A moth flies a distance of 3 m in only one-third of a second.
    a. What does the ratio 3/(1/3) tell you about the moth's motion? Explain.

    b. What does the ratio (1/3)/3 tell you about the moth's motion?

    c. How far would the moth fly in one-tenth of a second?

    d. How long does it take the moth to fly 4 m?

13. a. If someone drives at 25 miles per hour, is it necessary that he or she does so for an hour?

    b. Is it necessary to have a cubic centimeter of gold to say that gold has a density of 19.3 grams per cubic centimeter? Explain.

## 1.4 A Sense of Scale: Significant Figures, Scientific Notation, and Units

14. How many significant figures does each of the following numbers have?

a. 6.21 ________ e. 0.0621 ________ i. 1.0621 ________

b. 62.1 ________ f. 0.620 ________ j. $6.21 \times 10^3$ ________

c. 6210 ________ g. 0.62 ________ k. $6.21 \times 10^{-3}$ ________

d. 6210.0 ________ h. .62 ________ l. $62.1 \times 10^3$ ________

15. Compute the following numbers, applying the significant figure standards adopted for this text.

a. $33.3 \times 25.4 =$ ________ e. $2.345 \times 3.321 =$ ________

b. $33.3 - 25.4 =$ ________ f. $(4.32 \times 1.23) - 5.1 =$ ________

c. $33.3 \div 45.1 =$ ________ g. $33.3^2 =$ ________

d. $33.3 \times 45.1 =$ ________ h. $\sqrt{33.3} =$ ________

16. Express the following numbers and computed results in scientific notation, paying attention to significant figures.

a. 9,827 = ________ d. $32{,}014 \times 47 =$ ________

b. 0.000000550 = ________ e. $0.059 \div 2{,}304 =$ ________

c. 3,200,000 = ________ f. $320. \times 0.050 =$ ________

17. Convert the following to SI units. Work across the line and show all steps in the conversion. Use scientific notation and apply the proper use of significant figures. **Note:** Think carefully about g and h. Pictures may help.

a. 9.12 $\mu$s $\times$

b. 3.42 km $\times$

c. 44 cm/ms $\times$

d. 80 km/hr $\times$

e. 60 mph $\times$

f. 8 in $\times$

g. 14 in$^2$ $\times$

h. 250 cm$^3$ $\times$

18. Use Tables 1.4 and 1.5 and Examples 1.4 and 1.5 to assess whether or not the following statements are *reasonable*.

a. Joe is 180 cm tall.

b. I rode my bike to campus at a speed of 50 m/s.

c. A skier reaches the bottom of the hill going 25 m/s.

d. I can throw a ball a distance of 2 km.

e. I can throw a ball at a speed of 50 km/hr.

f. Joan's newborn baby has a mass of 33 kg.

g. A hummingbird has a mass of 3.3 g.

## 1.5 Vectors and Motion: A First Look

19. For the following motion diagrams, draw an arrow to indicate the displacement vector between the initial and final positions.

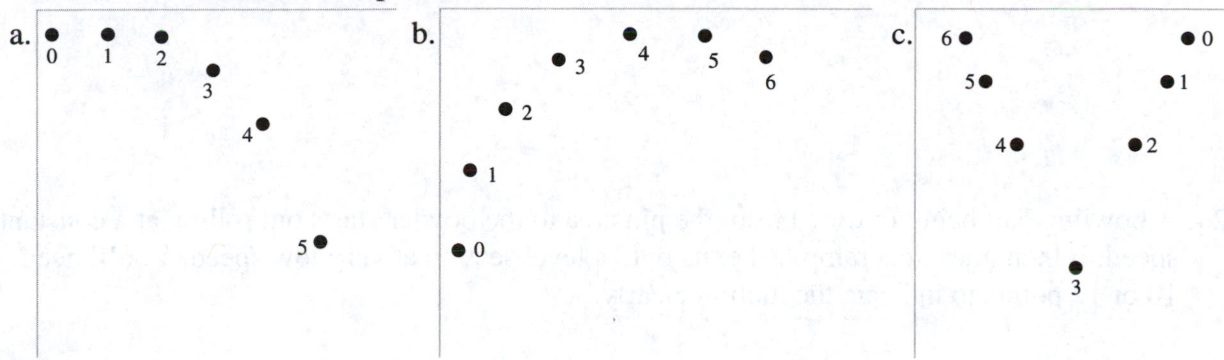


20. In each part of Exercise 19, is the object's displacement equal to the distance the object travels? Explain.

21. Draw and label the vector sum $\vec{A} + \vec{B}$.

a.

b.

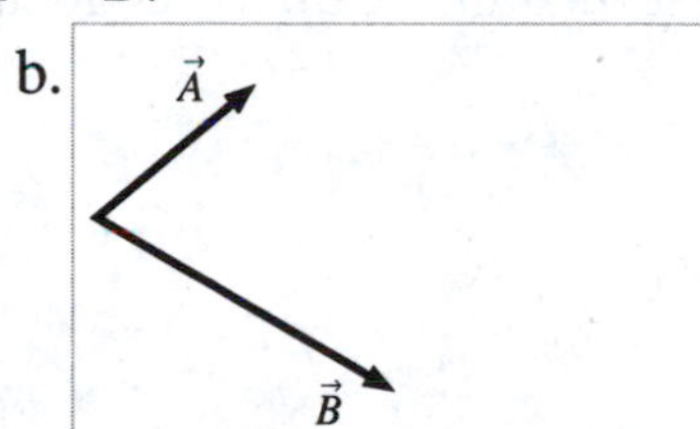


c.

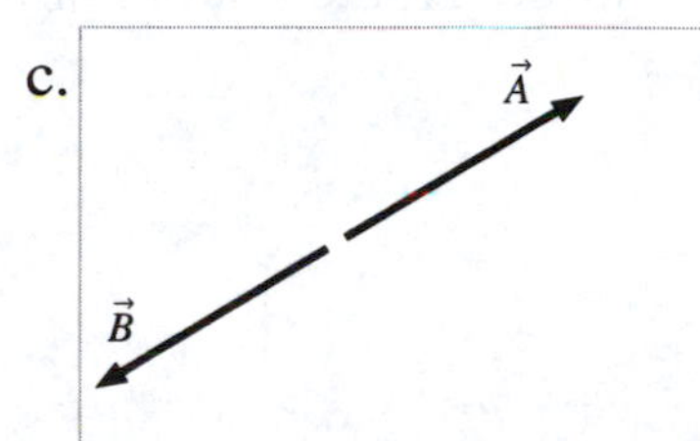


**Exercises 22–26:** Draw a motion diagram for each motion described below.
- Use the particle model.
- Show and label the *velocity* vectors.

22. Galileo drops a ball from the Leaning Tower of Pisa. Consider the ball's motion from the moment it leaves his hand until a microsecond before it hits the ground. Your diagram should be vertical.

23. A rocket-powered car on a test track accelerates from rest to a high speed, then coasts at constant speed after running out of fuel. Draw a vertical dashed line across your diagram to indicate the point at which the car runs out of fuel.

24. A bowling ball being returned from the pin area to the bowler starts out rolling at a constant speed. It then goes up a ramp and exits onto a level section at very low speed. You'll need 10 or 12 points to indicate the motion clearly.

25. A car is parked on a hill. The brakes fail, and the car rolls down the hill with an ever-increasing speed. At the bottom of the hill it runs into a thick hedge and gently comes to a halt.

26. Andy is standing on the street. Bob is standing on the second-floor balcony of their apartment, about 30 feet back from the street. Andy throws a baseball to Bob. Consider the ball's motion from the moment it leaves Andy's hand until a microsecond before Bob catches it.

# 2 Motion in One Dimension

## 2.1 Describing Motion

1. Sketch position-versus-time graphs for the following motions. Include a numerical scale on both axes with units that are *reasonable* for this motion. Some numerical information is given in the problems. For other quantities, make reasonable estimates.

   **Note:** A *sketched* graph is hand-drawn, rather than laid out with a ruler. Even so, a sketch must be neat, accurate, and include axis labels.

   a. A student walks to the bus stop, waits for the bus, and then rides to campus. Assume that all the motion is along a straight street.

   b. A student walks slowly to the bus stop, realizes he forgot his paper that is due, and *quickly* walks home to get it.

   c. The quarterback drops back 10 yards from the line of scrimmage, and then throws a pass 20 yards to a receiver, who catches it and sprints 20 yards to the goal. Draw your graph for the *football*. Think carefully about what the slopes of the lines should be.

2. The position-versus-time graph below shows the position of an object moving in a straight line for 12 seconds.

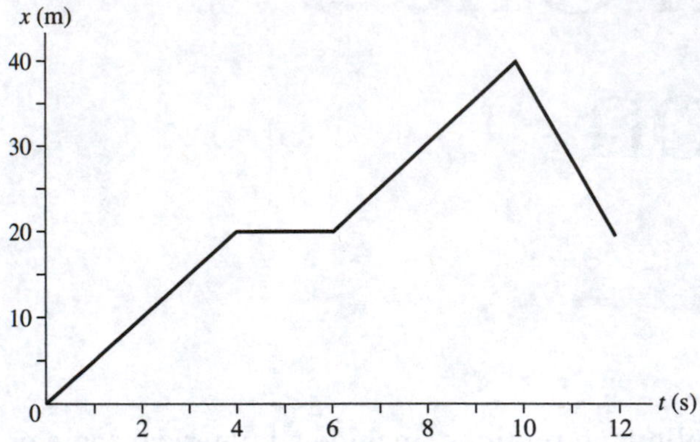


a. What is the position of the object at 2 s, 6 s, and 10 s after the start of the motion?

At 2 s: __________

At 6 s: __________

At 10 s: __________

b. What is the object's velocity during the first 4 s of motion?

c. What is the object's velocity during the interval from $t = 4$ s to $t = 6$ s?

d. What is the object's velocity during the four seconds from $t = 6$ s to $t = 10$ s?

e. What is the object's velocity during the final two seconds from $t = 10$ s to $t = 12$ s?

3. Interpret the following position-versus-time graphs by writing a very short "story" of what is happening. Be creative! Have characters and situations! Simply saying that "a car moves 100 meters to the right" doesn't qualify as a story. Your stories should make *specific reference* to information you obtain from the graphs, such as distances moved or time elapsed.

   a. Moving car

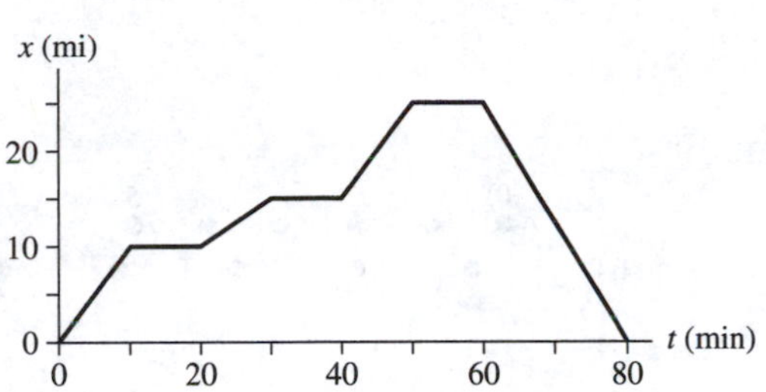


   b. Sprinter

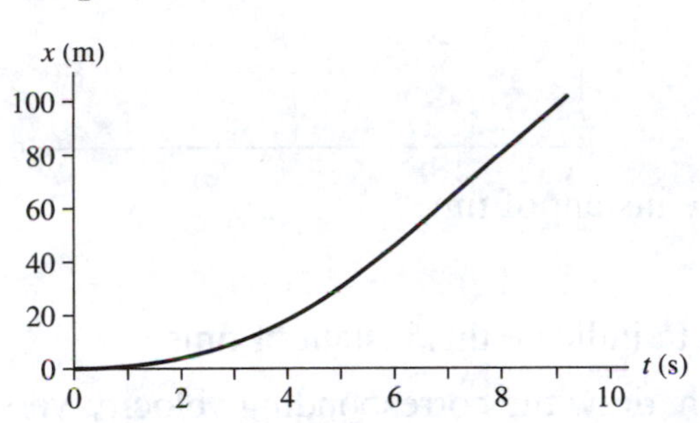


   c. Two football players

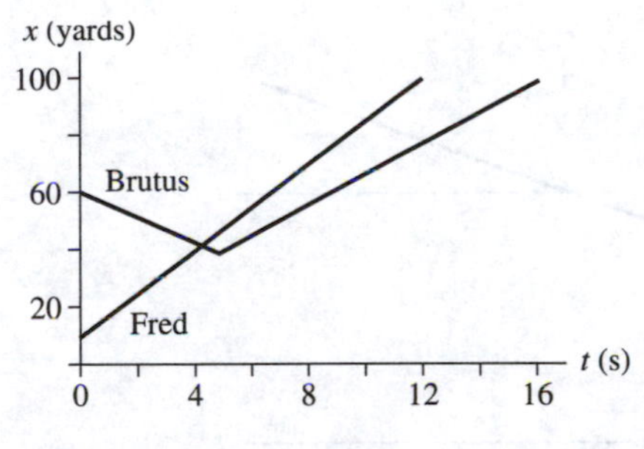


4. The figure shows a position-versus-time graph for the motion of objects A and B that are moving along the same axis.

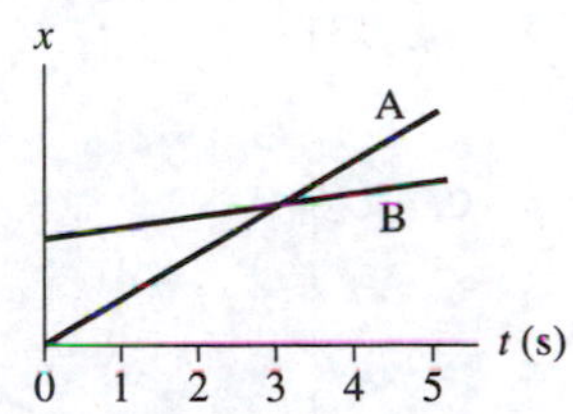


   a. At the instant $t = 1$ s, is the speed of A greater than, less than, or equal to the speed of B? Explain.

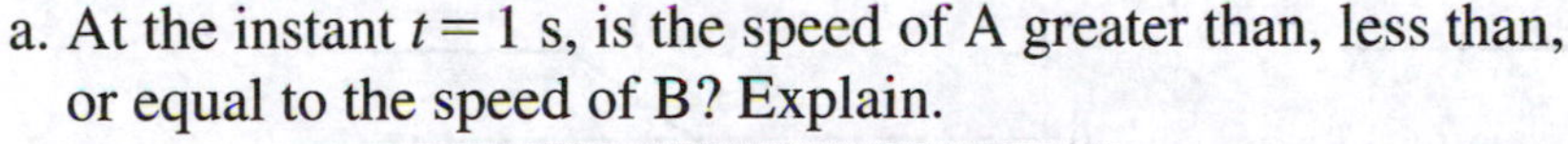

   b. Do objects A and B ever have the *same* speed? If so, at what time or times? Explain.

5. Draw both a position-versus-time graph *and* a velocity-versus-time graph for an object that is at rest at $x = 1$ m.

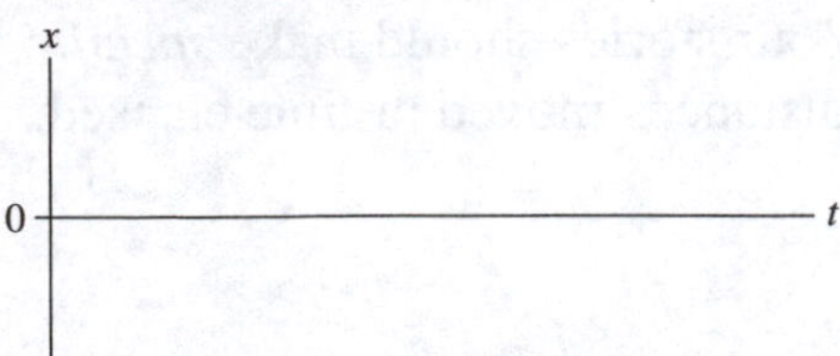


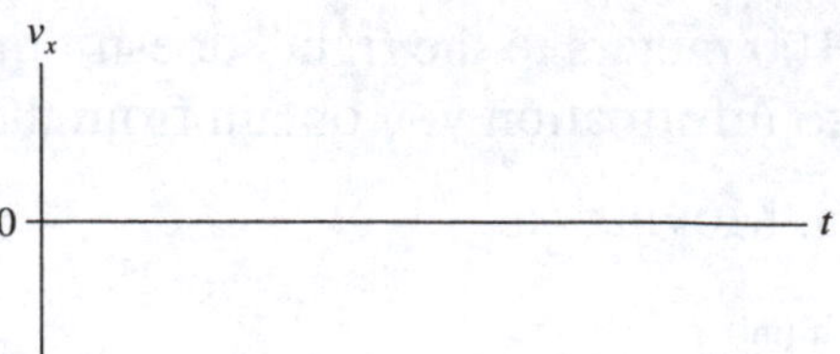


6. The figure shows six frames from the motion diagram of two moving cars, A and B.

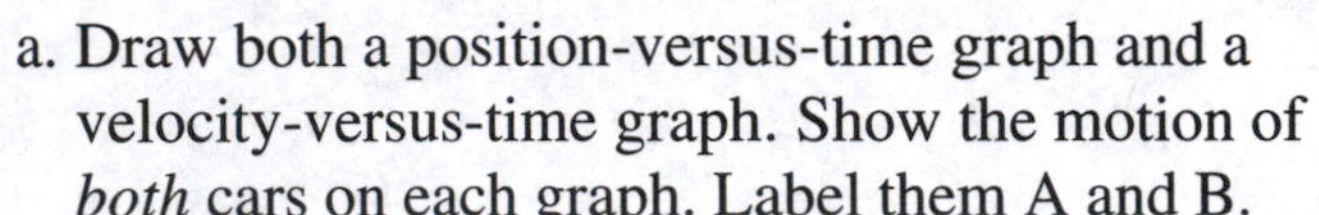

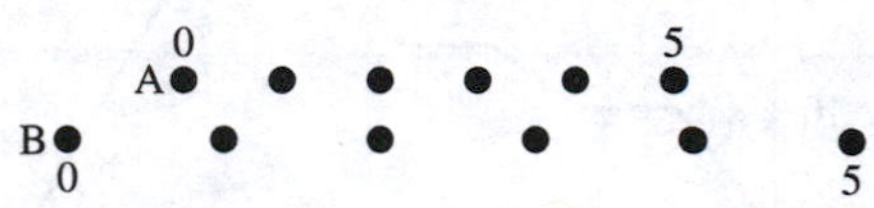


a. Draw both a position-versus-time graph and a velocity-versus-time graph. Show the motion of *both* cars on each graph. Label them A and B.

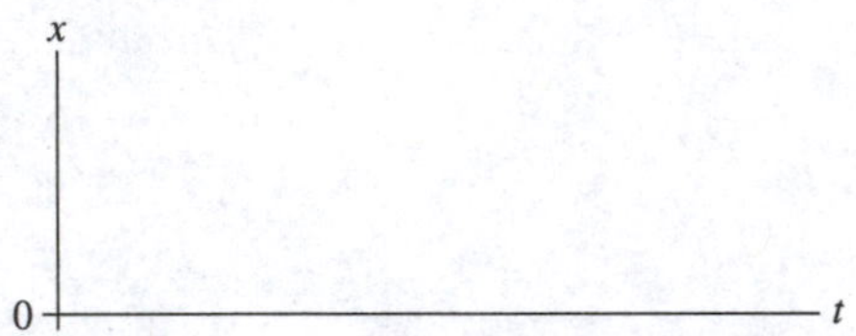

b. Do the two cars ever have the same position at one instant of time?

If so, in which frame number (or numbers)? ______________

Draw a vertical line through your graphs of part a to indicate this instant of time.

7. Below are four position-versus-time graphs. For each, draw the corresponding velocity-versus-time graph directly below it. A vertical line drawn through both graphs should connect the velocity $v_x$ at time $t$ with the position $x$ at the *same* time $t$. There are no numbers, but your graphs should correctly indicate the *relative* speeds.

a.

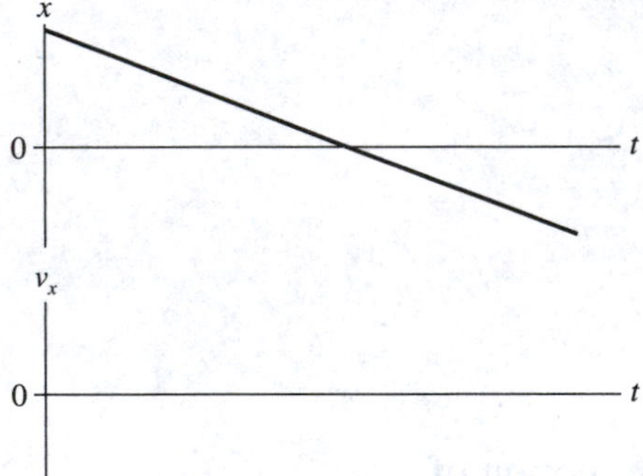


b.

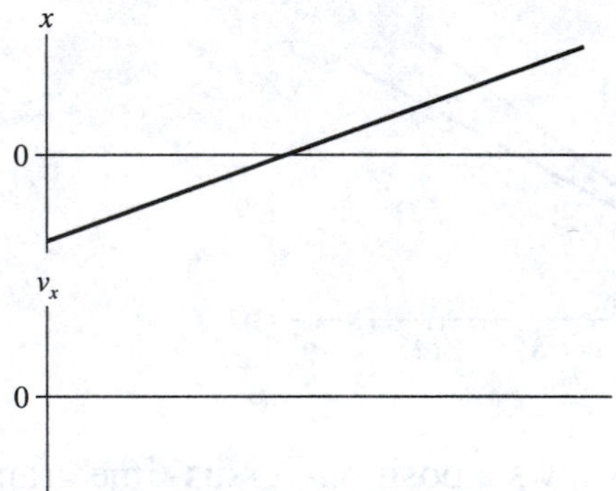


c.

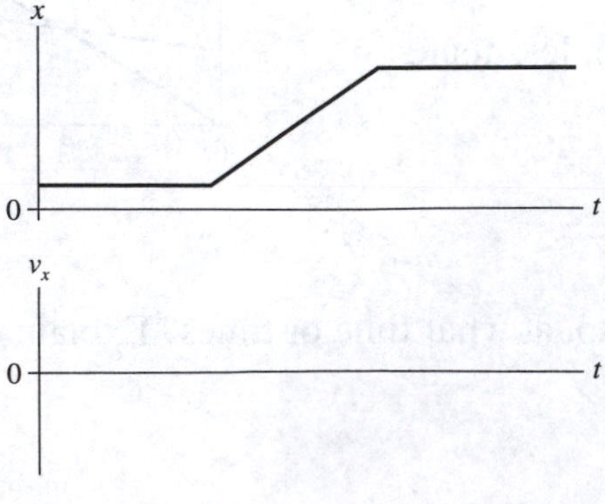


d.

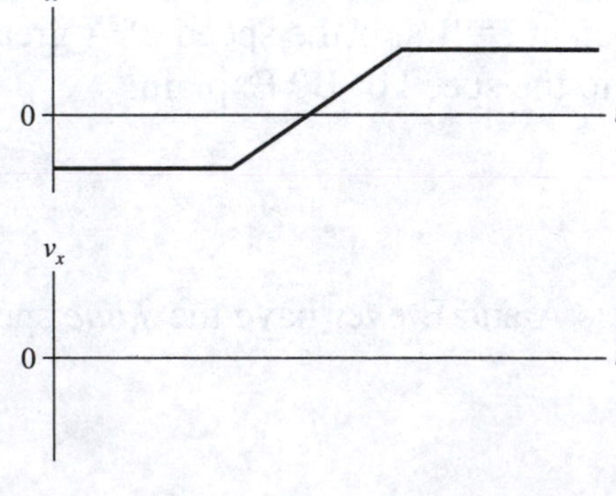

8. Below are two velocity-versus-time graphs. For each:
   - Draw the corresponding position-versus-time graph.
   - Give a written description of the motion.

   Assume that the motion takes place along a horizontal line and that $x_i = 0$.

a.

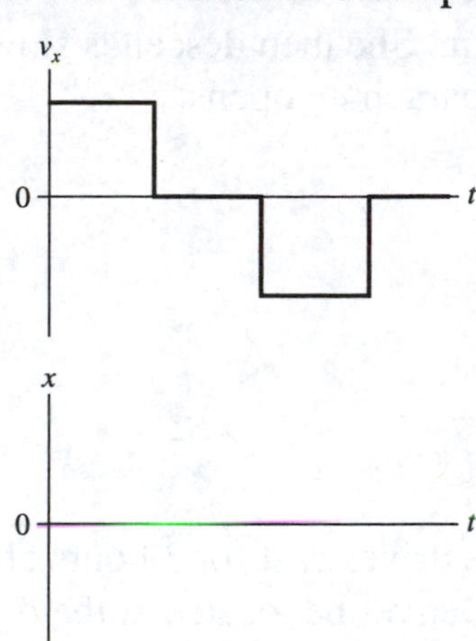


b.

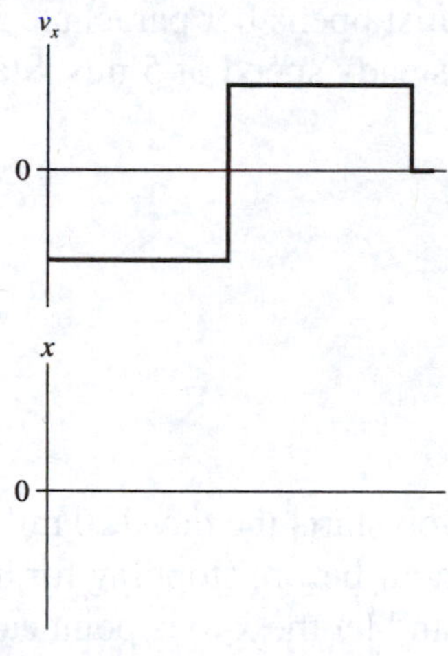


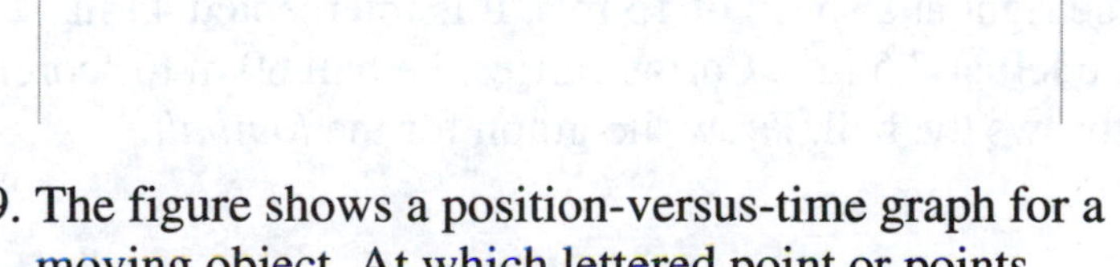

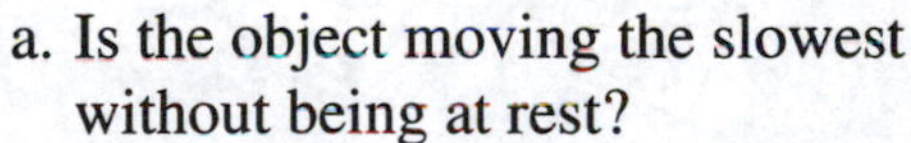

9. The figure shows a position-versus-time graph for a moving object. At which lettered point or points

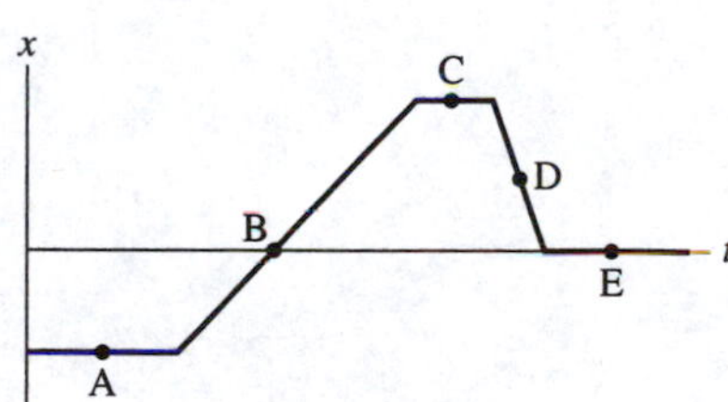


   a. Is the object moving the slowest without being at rest? ____________

   b. Is the object moving the fastest? ____________

   c. Is the object at rest? ____________

   d. Does the object have a constant nonzero velocity? ____________

   e. Is the object moving to the left? ____________

## 2.2 Uniform Motion

10. Sketch position-versus-time graphs for the following motions. Include appropriate numerical scales along both axes. A small amount of computation may be necessary.
    a. A parachutist opens her parachute at an altitude of 1500 m. She then descends slowly to earth at a steady speed of 5 m/s. Start your graph as her parachute opens.

    b. Trucker Bob starts the day 120 miles west of Denver. He drives east for 3 hours at a steady 60 miles/hour before stopping for his coffee break. Let Denver be located at the origin, $x = 0$ mi, and let the $x$-axis point east.

    c. Quarterback Bill throws the ball to the right at a speed of 15 m/s. It is intercepted 45 m away by Carlos, who is running to the left at 7.5 m/s. Carlos carries the ball 60 m to score. Let $x = 0$ m be the point where Bill throws the ball. Draw the graph for the *football*.

11. The height of a building is proportional to the number of stories it has. If a 25-story building is 260 ft tall, what is the height of an 80-story building? Answer this using ratios, not by calculating the height per story.

## 2.3 Instantaneous Velocity

12. Below are two position-versus-time graphs. For each, draw the corresponding velocity-versus-time graph directly below it. A vertical line drawn through both graphs should connect the velocity $v_x$ at time $t$ with the position $x$ at the *same* time $t$. There are no numbers, but your graphs should correctly indicate the *relative* speeds.

a.

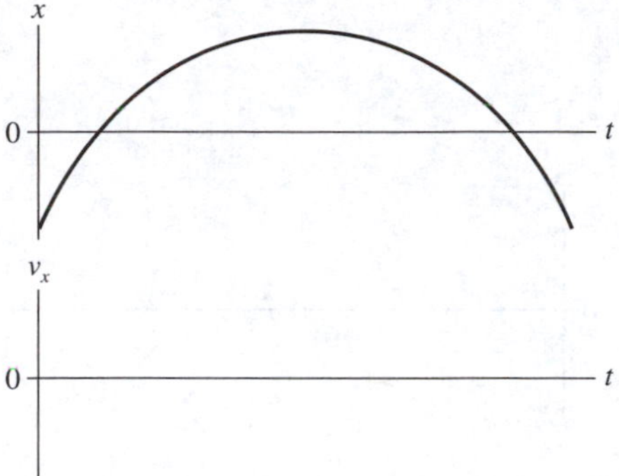

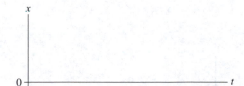

b.

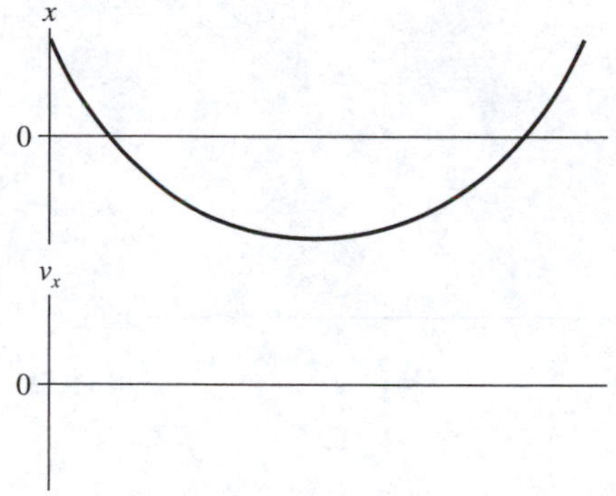

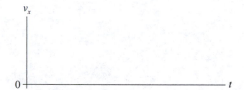

13. The figure shows six frames from the motion diagram of two moving cars, A and B.

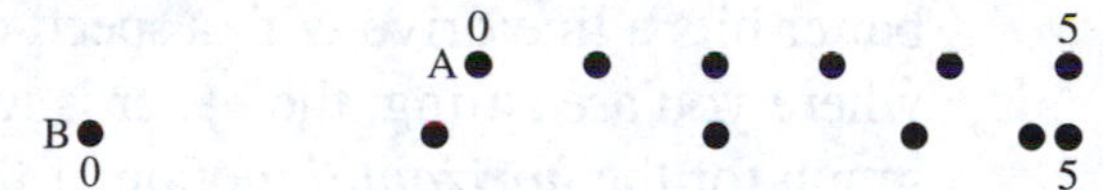


a. Draw both a position-versus-time graph and a velocity-versus-time graph. Show *both* cars on each graph. Label them A and B.

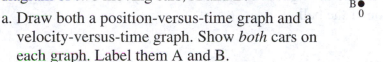

b. Do the two cars ever have the same position at one instant of time? __________

If so, in which frame number (or numbers)? __________

Draw a vertical line through your graphs of part a to indicate this instant of time.

c. Do the two cars ever have the same velocity at one instant of time? __________

If so, between which two frames? __________

14. For each of the following motions, draw
  - A motion diagram and
  - Both position and velocity graphs.

  a. A car starts from rest, steadily speeds up to 40 mph in 15 s, moves at a constant speed for 30 s, and then comes to a halt in 5 s.

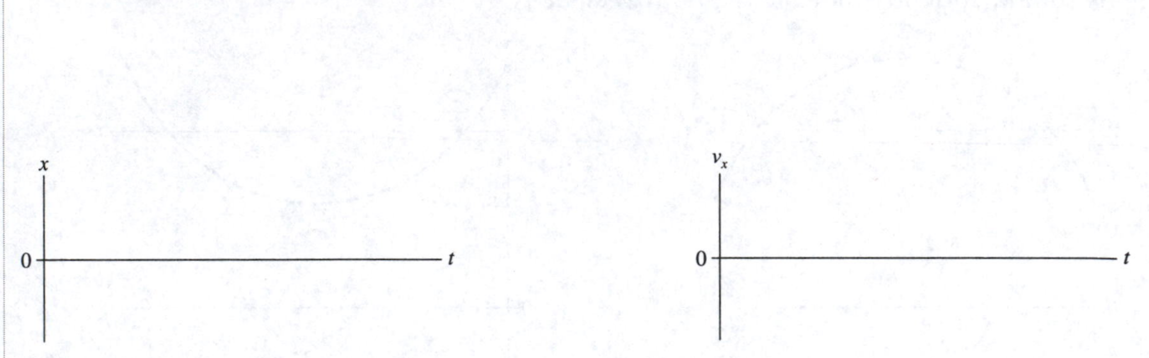


  b. A pitcher winds up and throws a baseball with a speed of 40 m/s. One-half second later, the batter hits a line drive with a speed of 60 m/s. The ball is caught 1 s after it is hit. From where you are sitting, the batter is to the right of the pitcher. Draw your motion diagram and graph for the *horizontal* motion of the ball.

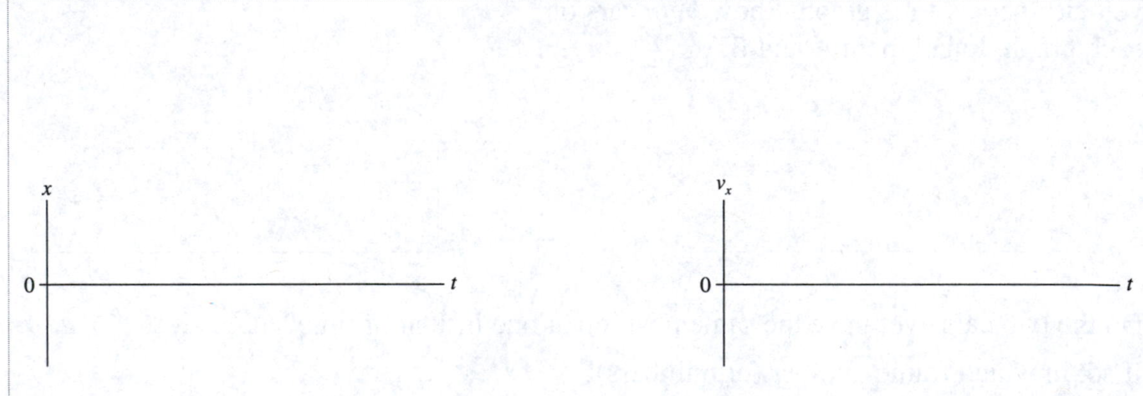


15. Below are three velocity-versus-time graphs. For each, draw the corresponding position-versus-time graph. Assume that the motion is horizontal and that $x_i = 0$.

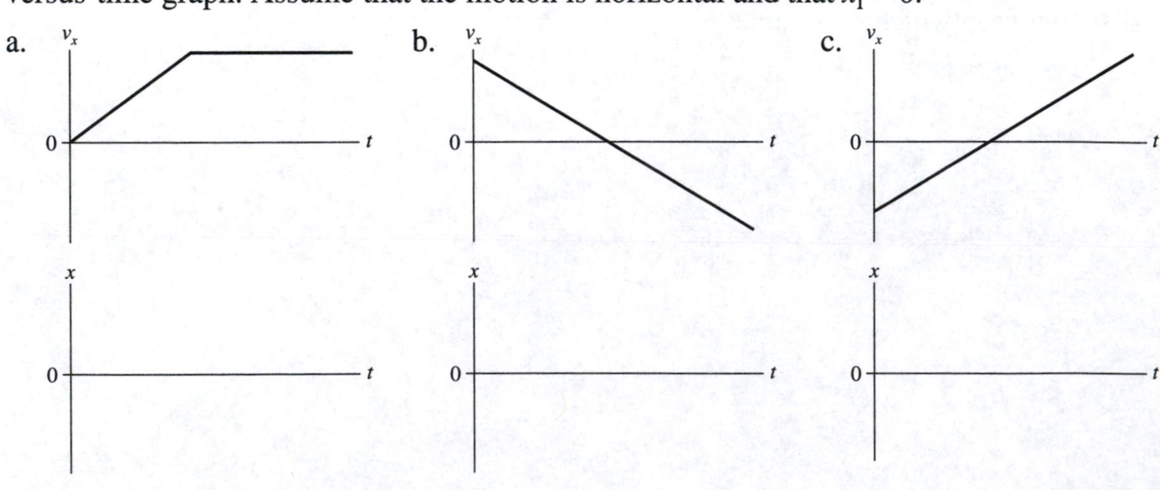

## 2.4 Acceleration

16. The four motion diagrams below show an initial point 0 and a final point 1. A pictorial representation would define the five symbols: $x_0$, $x_1$, $v_{0x}$, $v_{1x}$, and $a_x$ for horizontal motion and equivalent symbols with $y$ for vertical motion. Determine whether each of these quantities is positive, negative, or zero. Give your answer by writing +, –, or 0 in the table below.

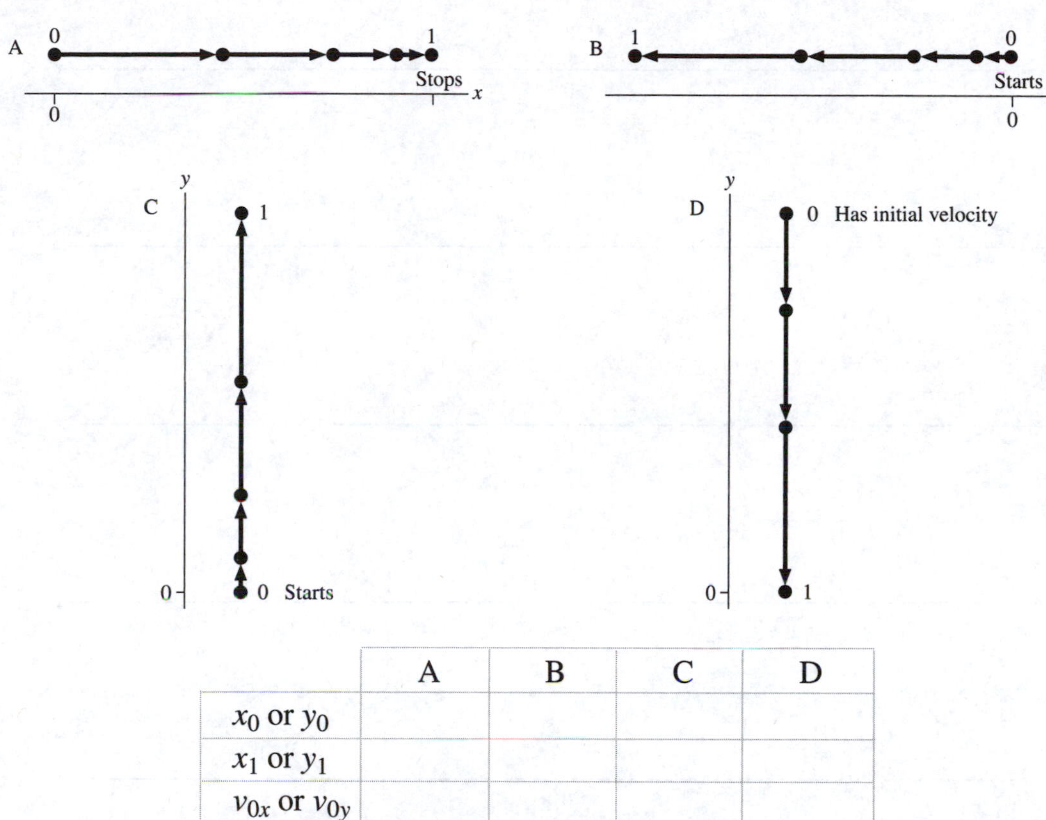


| | A | B | C | D |
|---|---|---|---|---|
| $x_0$ or $y_0$ | | | | |
| $x_1$ or $y_1$ | | | | |
| $v_{0x}$ or $v_{0y}$ | | | | |
| $v_{1x}$ or $v_{1x}$ | | | | |
| $a_x$ or $a_y$ | | | | |

17. The three symbols $x$, $v_x$, and $a_x$ have eight possible combinations of *signs*. For example, one combination is $(x, v_x, a_x) = (+, –, +)$.

a. List all eight combinations of signs for $x$, $v_x$, $a_x$.

1. ________ 5. ________

2. ________ 6. ________

3. ________ 7. ________

4. ________ 8. ________

b. For each of the eight combinations of signs you identified in part a:

- Draw a four-dot motion diagram of an object that has these signs for $x$, $v_x$, and $a_x$.
- Draw the diagram *above* the axis whose number corresponds to part a.
- Use **black** and **red** for your $\vec{v}$ and $\vec{a}$ vectors. Be sure to label the vectors.

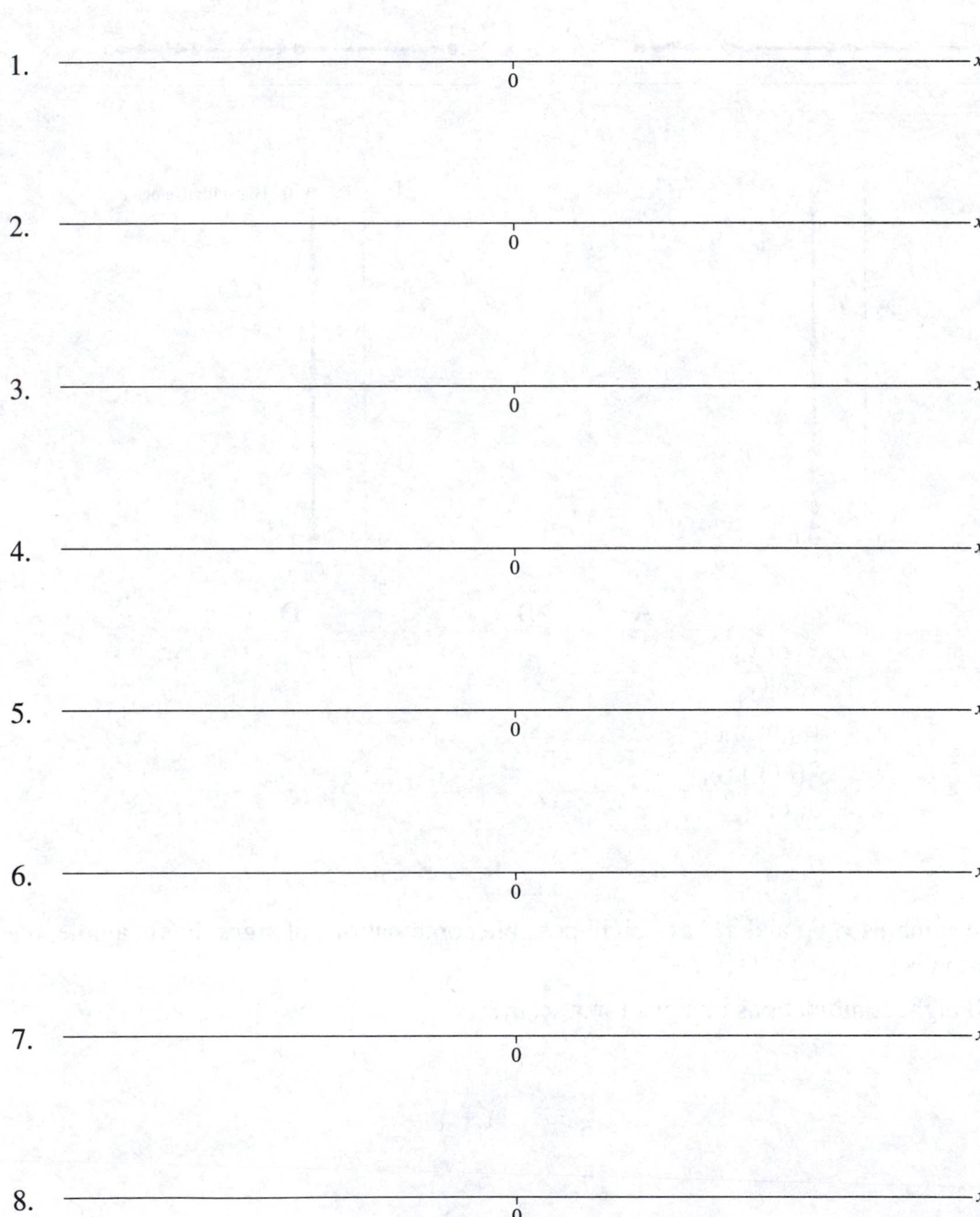

## 2.5 Motion with Constant Acceleration

18. Draw a motion diagram to illustrate each of the following situations.

a. $a_x = 0$ but $v_x \neq 0$.

b. $v_x = 0$ but $a_x \neq 0$.

c. $v_x < 0$ and $a_x > 0$.

19. The quantity $y$ is proportional to the square of $x$, and $y = 36$ when $x = 3$.

a. Find $y$ if $x = 5$. ______

b. Find $x$ if $y = 16$. ______

c. By what factor must $x$ change for the value of $y$ to double? ______

d. Consider the equation in your text relating $\Delta x$ and $\Delta t$ for motion with constant acceleration $a_x$. Which of these three quantities plays the role of $x$ in a quadratic relationship $y = Ax^2$? Which plays the role of $y$?

20. Below are three velocity-versus-time graphs. For each, draw the corresponding acceleration-versus-time graph. Then draw a motion diagram below the graphs.

a.

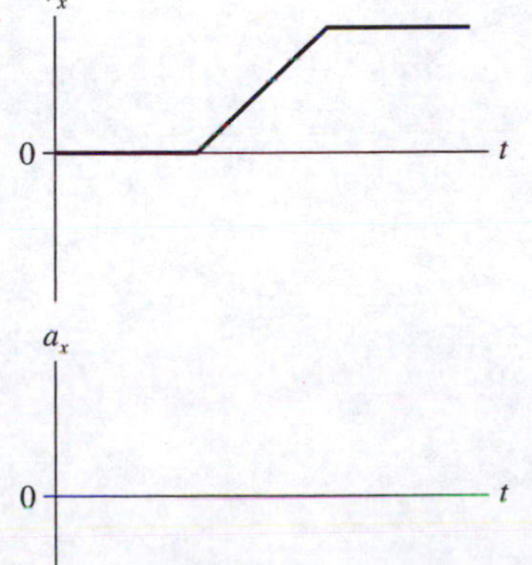


b.

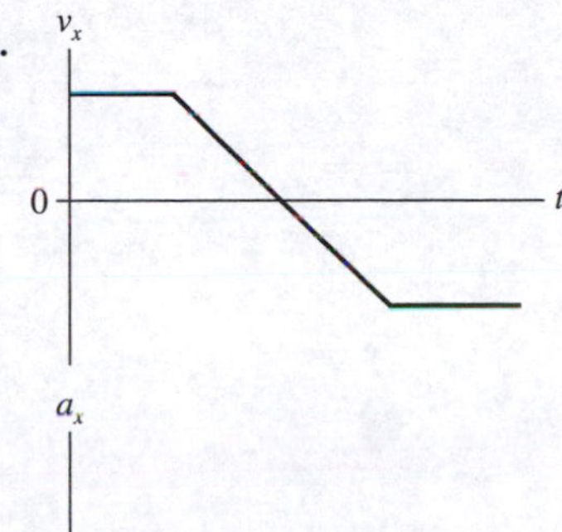


c.

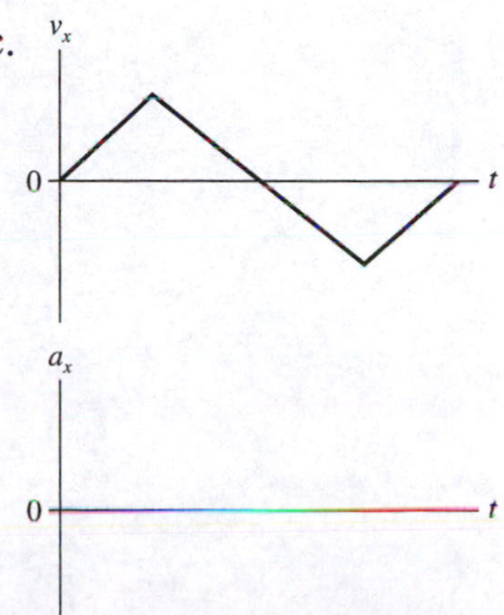

## 2.6 Solving One-Dimensional Motion Problems

21. Draw a pictorial representation of each situation described below. That is, (i) sketch the situation, showing appropriate points in the motion, (ii) establish a coordinate system on your sketch, and (iii) define appropriate symbols for the known and unknown quantities. **Do not solve.** See textbook Figure 2.31 as an example.

a. A car traveling at 30 m/s screeches to a halt, leaving 55-m-long skid marks. What was the car's acceleration while braking?

b. A bicyclist starts from rest and accelerates at 4.0 $m/s^2$ for 3.0 s. The cyclist then travels for 20 s at a constant speed. How far does the cyclist travel?

c. You are driving your car at 12 m/s when a deer jumps in front of your car. What is the shortest stopping distance for your car if your reaction time is 0.80 s and your car brakes at 6.0 $m/s^2$?

## 2.7 Free Fall

22. A ball is thrown straight up into the air. At each of the following instants, is the ball's acceleration $g$, $-g$, 0, $< g$, or $> g$?
    a. Just after leaving your hand? ____________
    b. At the very top (maximum height)? ____________
    c. Just before hitting the ground? ____________

23. A ball is thrown straight up into the air. It reaches height $h$, and then falls back down to the ground. On the axes below, graph the ball's position, velocity, and acceleration from an instant after it leaves the thrower's hand until an instant before it hits the ground. Indicate on your graphs the times during which the ball is moving upward, at its peak, and moving downward.

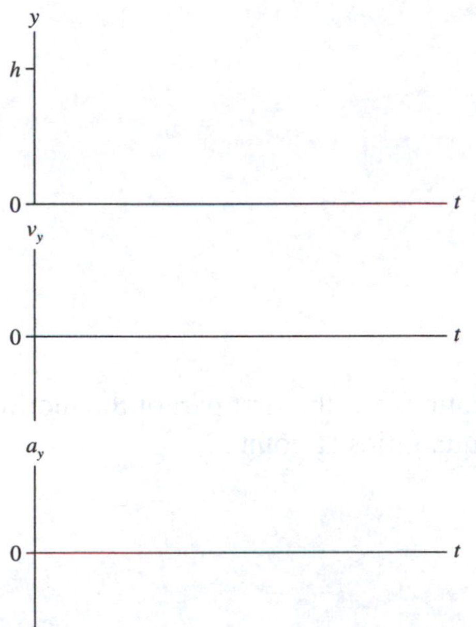


24. A rock is *thrown* (not dropped) straight down from a bridge into the river below.
    a. Immediately *after* being released, is the magnitude of the rock's acceleration greater than $g$, less than $g$, or equal to $g$? Explain.

    b. Immediately before hitting the water, is the magnitude of the rock's acceleration greater than $g$, less than $g$, or equal to $g$? Explain.

25. A model rocket is launched straight up with constant acceleration *a*. It runs out of fuel at time *t*.
PSS 2.1 Suppose you need to determine the maximum height reached by the rocket. We'll assume that air resistance is negligible.
   a. Is the rocket at maximum height the instant it runs out of fuel? ____________
   b. Is there anything other than gravity acting on the rocket after it runs out of fuel? ________
   c. What is the name of motion under the influence of only gravity? ______________
   d. Draw a pictorial representation for this problem. You should have three identified points in the motion: launch, out of fuel, maximum height. Call these points 1, 2, and 3.
      - Using subscripts, define 11 quantities: $y$, $v_y$, and $t$ at each of the three points, plus acceleration $a_1$ connecting points 1 and 2 and acceleration $a_2$ connecting points 2 and 3.
      - 7 of these quantities are Knowns; identify them and specify their values. Some are 0. Others can be given in terms of $a$, $t$, and $g$, which are "known," because they're stated in the problem, even though you don't have numerical values for them. For example, $t_1 = t$. Be careful with signs!
      - Identify which one of the 4 unknown quantities you're trying to find.
   e. This is a two-part problem. Write two kinematic equations for the first part of the motion to determine—again symbolically—the two unknown quantities at point 2.

   f. Now write a kinematic equation for the second half of the motion that will allow you to find the desired unknown that will answer the question. Just write the equation; don't yet solve it.

   g. Now, substitute what you learned in part e into your equation of part f, do the algebra to solve for the unknown, and simplify the result as much as possible.

## You Write the Problem!

**Exercises 26–28:** You are given the kinematic equation that is used to solve a problem. For each of these:

a. Write a *realistic* physics problem for which this is the correct equation. Look at worked examples and end-of-chapter problems in the textbook to see what realistic physics problems are like. Be sure that the problem you write, and the answer you ask for, is consistent with the information given in the equation.
b. Draw the motion diagram and pictorial representation for your problem.
c. Finish the solution of the problem.

26. $64\text{ m} = 0.0\text{ m} + (32\text{ m/s})(4.0\text{ s} - 0.0\text{ s}) + \frac{1}{2}a_x(4.0\text{ s} - 0.0\text{ s})^2$

27. $0.0\text{ m/s} = 36\text{ m/s} - (3.0\text{ m/s}^2)t$

28. $(10\text{ m/s})^2 = (v_y)_i^2 - 2(9.8\text{ m/s}^2)(10\text{ m} - 0\text{ m})$

# 3 Vectors and Motion in Two Dimensions

## 3.1 Using Vectors

**Exercises 1–3:** Draw and label the vector sum $\vec{A} + \vec{B}$ or $\vec{A} + \vec{B} + \vec{C}$.

1.

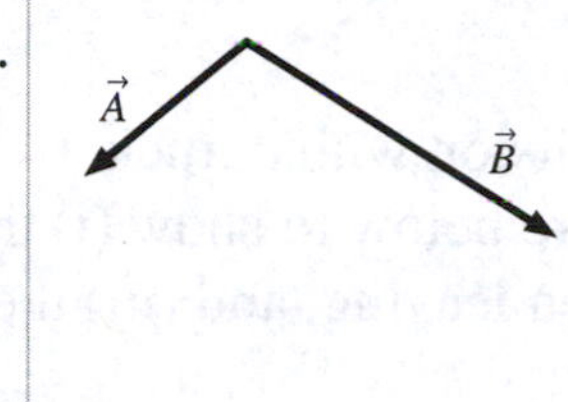


2.

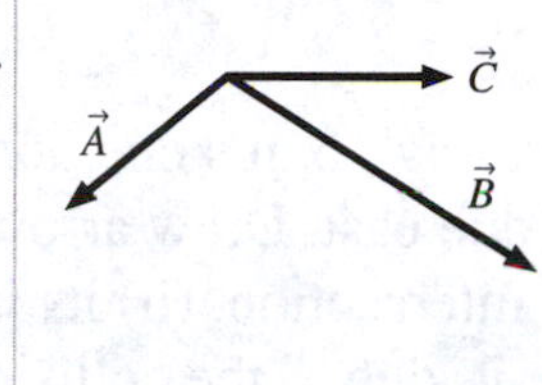


3.

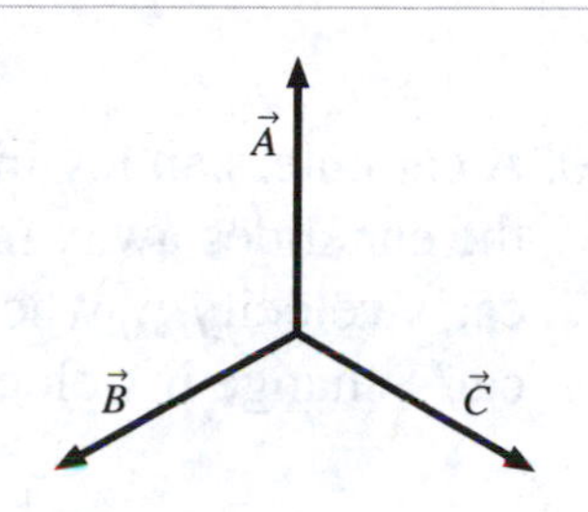


4. Draw and label the vector $2\vec{A}$ and the vector $\frac{1}{2}\vec{A}$.

**Exercises 5–7:** Draw and label the vector difference $\vec{A} - \vec{B}$.

5.

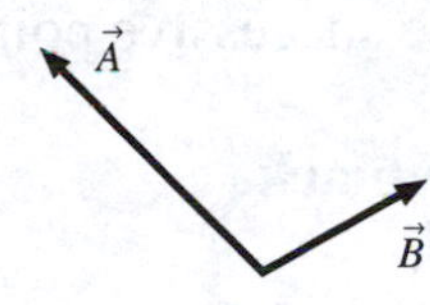


6.

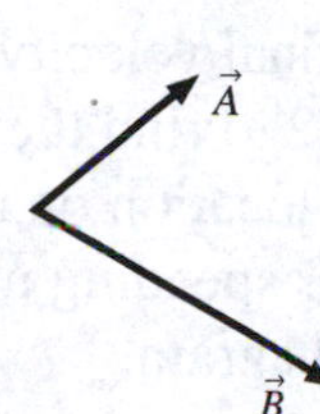


7.

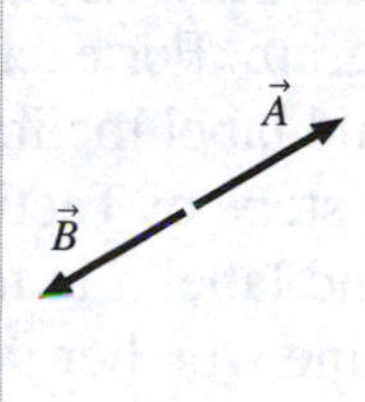


8. Given vectors $\vec{A}$ and $\vec{B}$ below, find the vector $\vec{C} = 2\vec{A} - 3\vec{B}$.

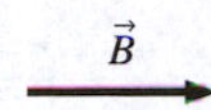

## 3.2 Using Vectors on Motion Diagrams

9. The figure below shows the positions of a moving object at three successive points in a motion diagram. Draw and label the velocity vector $\vec{v}_0$ for the motion from 0 to 1 and the vector $\vec{v}_1$ for the motion from 1 to 2. Then determine and draw the vector $\vec{v}_1 - \vec{v}_0$ with its tail on point 1.

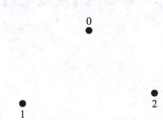

10. A car enters an icy intersection traveling 16 m/s due north. After a collision with a truck, the car slides away moving 12 m/s due east. Draw arrows on the picture below to show (i) the car's velocity $\vec{v}_0$ when entering the intersection, (ii) its velocity $\vec{v}_1$ when leaving, and (iii) the car's change in velocity $\Delta\vec{v} = \vec{v}_1 - \vec{v}_0$ due to the collision.

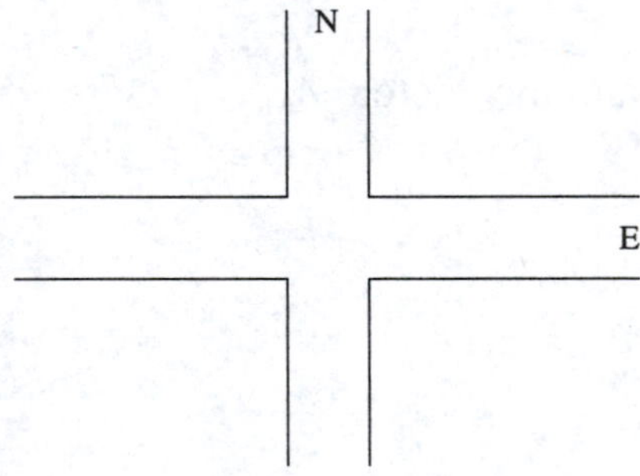


**Exercises 11–12:** The figures below show an object's position at three successive points in a motion diagram. For each diagram:

- Draw and label the initial and final velocity vectors $\vec{v}_0$ and $\vec{v}_1$. Use **black**.
- Use the steps of Tactics Box 3.2 to find the change in velocity $\Delta\vec{v}$.
- Draw and label $\vec{a}$ at the proper location on the motion diagram. Use **red**.
- Determine whether the object is speeding up, slowing down, or moving at a constant speed. Write your answer beside the diagram.

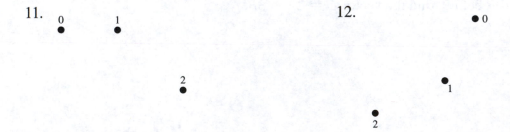

## 3.3 Coordinate Systems and Vector Components

**Exercises 13–15:** Draw and label the *x*- and *y*-component vectors of the vector shown.

13.

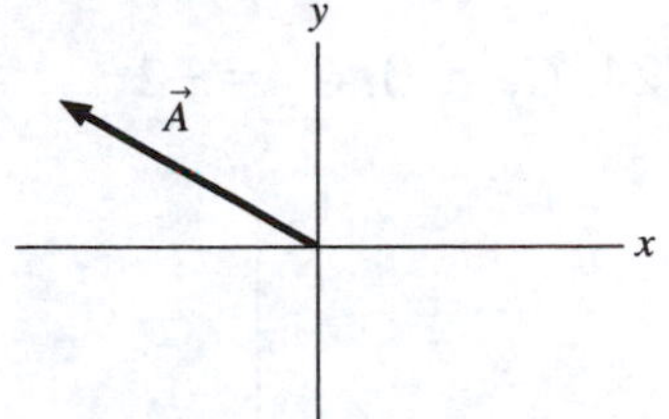


14.

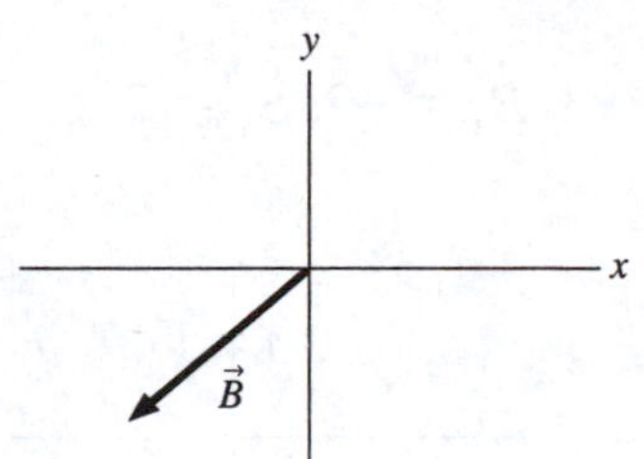


15.

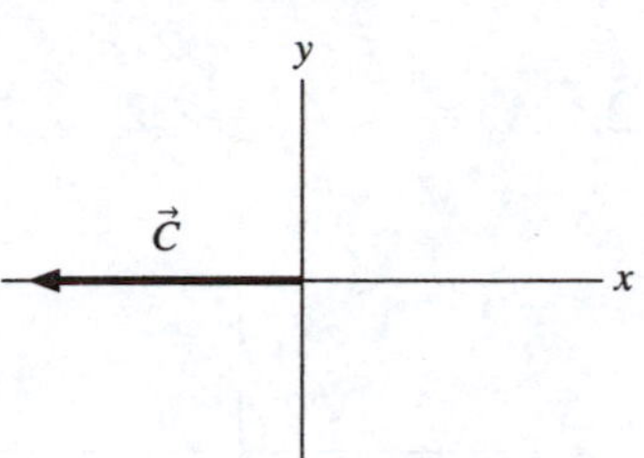


**Exercises 16–18:** Determine the numerical values of the *x*- and *y*-components of each vector.

16.

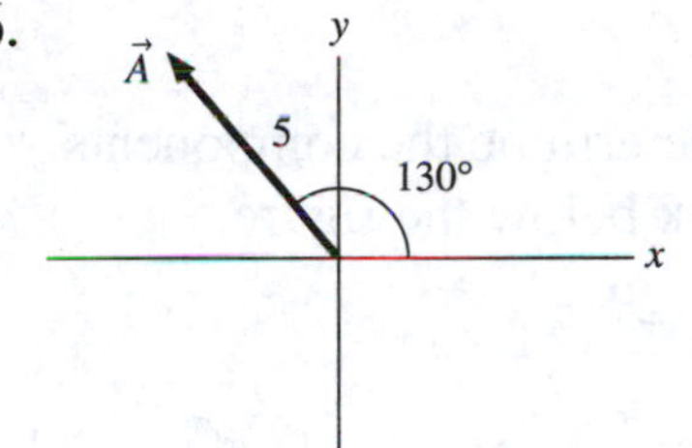


17.

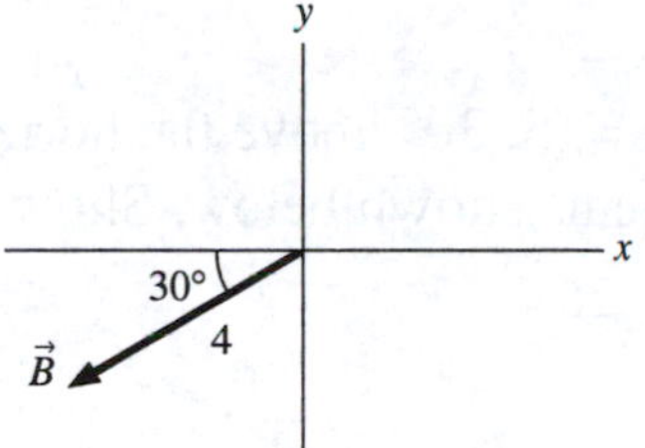


18.

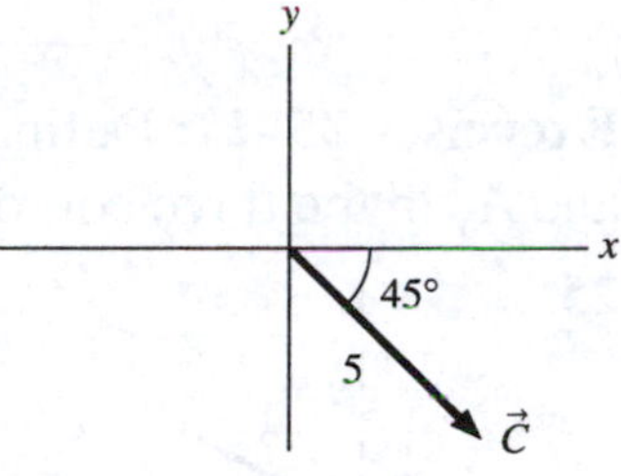


$A_x =$ ______________ $B_x =$ ______________ $C_x =$ ______________

$A_y =$ ______________ $B_y =$ ______________ $C_y =$ ______________

19. What is the vector sum $\vec{D} = \vec{A} + \vec{B} + \vec{C}$ of the three vectors defined in Exercises 16–18?

$D_x =$ ______________________ $D_y =$ ______________________

20. Can a vector have a component equal to zero and still have nonzero magnitude? Explain.

21. Can a vector have zero magnitude if one of its components is nonzero? Explain.

**Exercises 22–24:** For each vector:

- Draw the vector on the axes provided.
- Draw and label an angle $\theta$ to describe the direction of the vector.
- Find the magnitude and the angle of the vector.

22. $A_x = 3, A_y = -2$

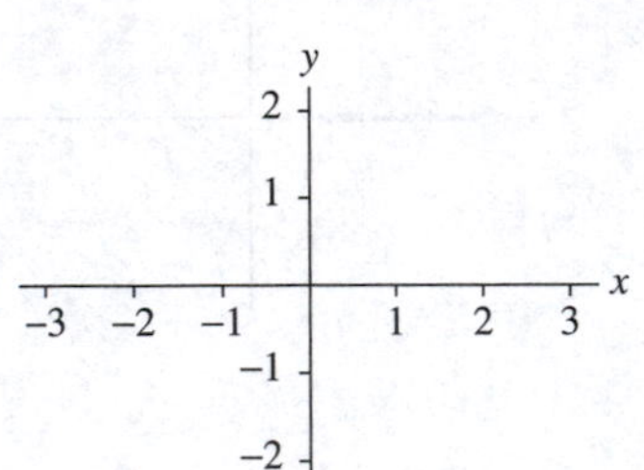


$A =$ ____________

$\theta =$ ____________

23. $B_x = -2, B_y = 2$

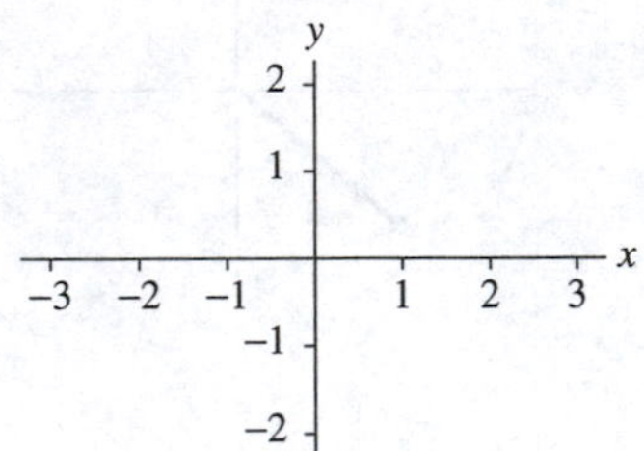


$B =$ ____________

$\theta =$ ____________

24. $C_x = 0, C_y = -2$

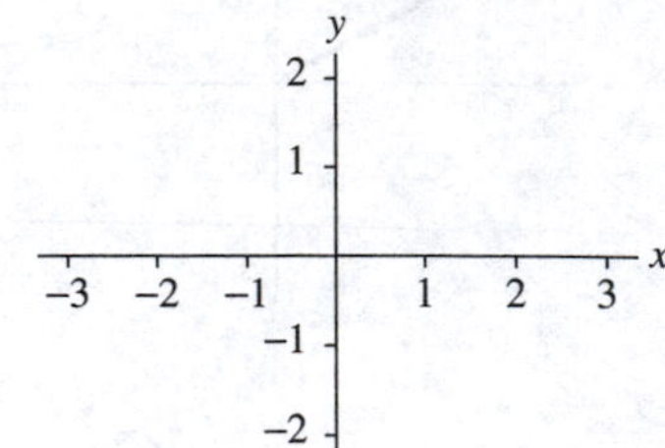


$C =$ ____________

$\theta =$ ____________

**Exercises 25–27:** Define vector $\vec{A}$ = (5, 30° above the horizontal). Determine the components $A_x$ and $A_y$ in the three coordinate systems shown below. Show your work below the figure.

25.

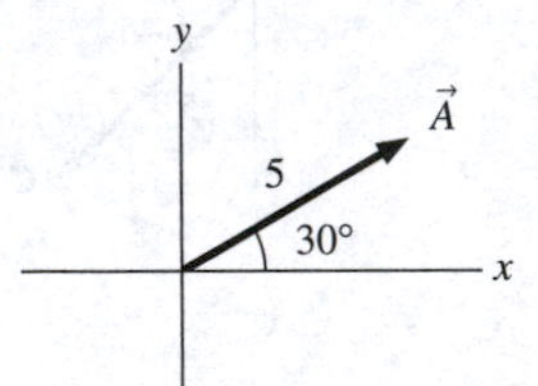


$A_x =$ ____________

$A_y =$ ____________

26.

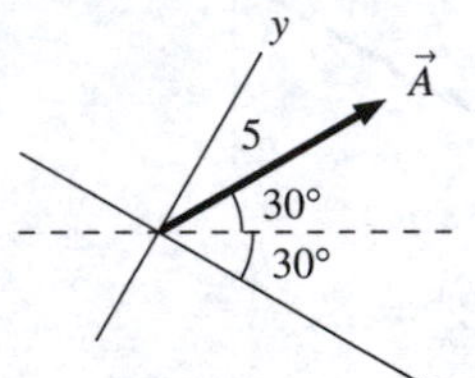


$A_x =$ ____________

$A_y =$ ____________

27.

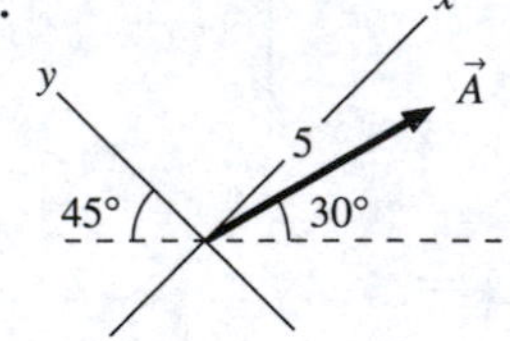


$A_x =$ ____________

$A_y =$ ____________

## 3.4 Motion on a Ramp

28. The figure shows a ramp and a ball that rolls along the ramp. Draw vector arrows on the figure to show the ball's acceleration at each of the lettered points A to E (or write $\vec{a} = \vec{0}$, if appropriate).

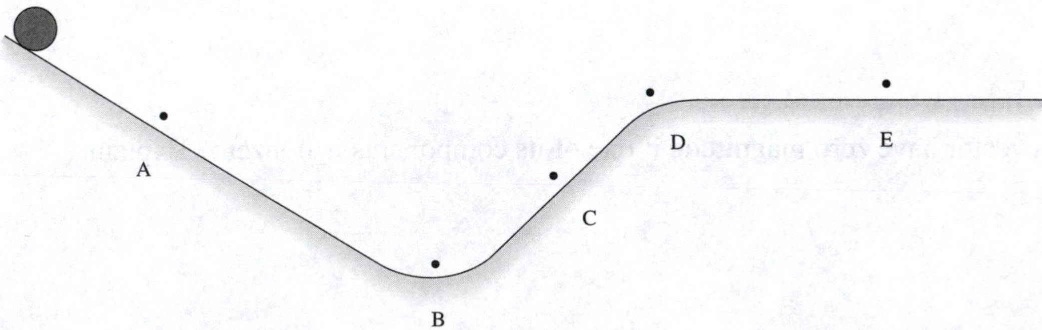

## 3.5 Relative Motion

29. On a ferry moving steadily forward in still water at 5 m/s, a passenger walks toward the back of the boat at 2 m/s. (i) Write a symbolic equation to find the velocity of the passenger with respect to the water using $(v_x)_{AB}$ notation. (ii) Substitute the appropriate values into your equation to determine the value of that velocity.

30. A boat crossing a river can move at 5 m/s with respect to the water. The river is flowing to the right at 3 m/s. In (a), the boat points straight across the river and is carried downstream by the water. In (b), the boat is angled upstream by the amount needed for it to travel straight across the river. For each situation, draw the velocity vectors $\vec{v}_{RS}$ of the river with respect to the shore, $\vec{v}_{BR}$ of the boat with respect to the river, and $\vec{v}_{BS}$ of the boat with respect to the shore.

a.

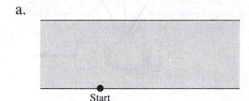

b.

Finish

Start

31. Ryan, Samantha, and Tomas are driving their convertibles. At the same instant, they each see a jet plane with an instantaneous velocity of 200 m/s and an acceleration of 5 m/s$^2$. Rank in order, from largest to smallest, the jet's *speed* $v_R$, $v_S$, and $v_T$ according to Ryan, Samantha, and Tomas. Explain.

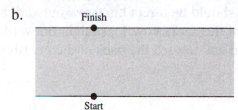

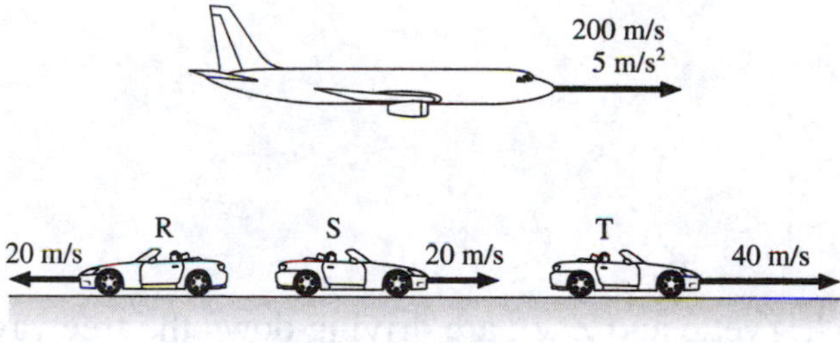


Order:

Explanation:

32. An electromagnet on the ceiling of an airplane holds a steel ball. When a button is pushed, the magnet releases the ball. The experiment is first done while the plane is parked on the ground, and the point where the ball hits the floor is marked with an X. Then the experiment is repeated while the plane is flying level at a steady 500 mph. Does the ball land slightly in front of the X (toward the nose of the plane), on the X, or slightly behind the X (toward the tail of the plane)? Explain.

33. Zack is driving past his house. He wants to toss his physics book out the window and have it land in his driveway. If he lets go of the book exactly as he passes the end of the driveway, should he direct his throw outward and toward the front of the car (throw 1), straight outward (throw 2), or outward and toward the back of the car (throw 3)? Explain.

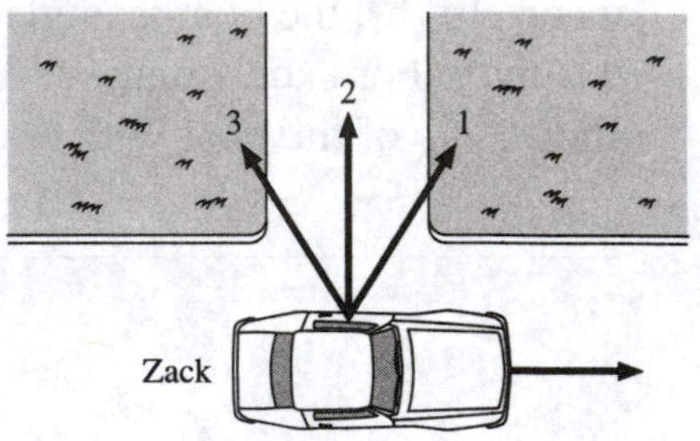


34. Yvette and Zack are driving down the freeway side by side with their windows rolled down. Zack wants to toss his physics book out the window and have it land in Yvette's front seat. Should he direct his throw outward and toward the front of the car (throw 1), straight outward (throw 2), or outward and toward the back of the car (throw 3)? Explain.

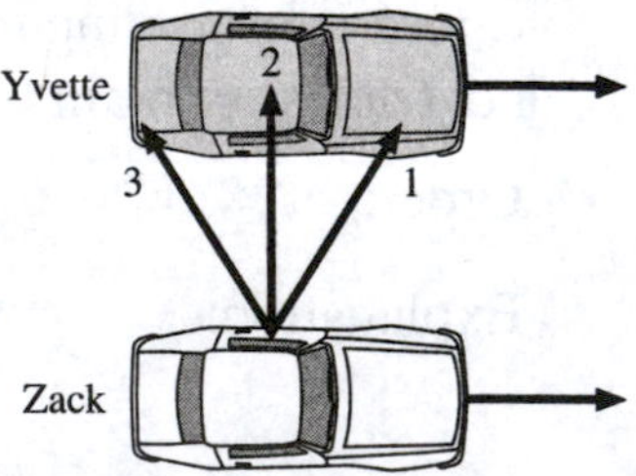

## 3.6 Motion in Two Dimensions: Projectile Motion

35. Complete the motion diagram for this trajectory, showing velocity and acceleration vectors.

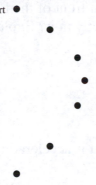

36. A projectile is launched over level ground and lands some distance away.

a. Is there any point on the trajectory where $\vec{v}$ and $\vec{a}$ are parallel to each other? If so, where?

b. Is there any point where $\vec{v}$ and $\vec{a}$ are perpendicular to each other? If so, where?

c. Which of the following remain constant throughout the flight: $x$, $y$, $v$, $v_x$, $v_y$, $a_x$, $a_y$?

37. A ball is projected horizontally at 10 m/s and hits the ground 2.0 s later. The figure shows the ball's position every 0.5 s.

- At each dot, starting with $t = 0.5$ s, draw a vector for the horizontal component of velocity $v_x$ and a vector for the vertical component of velocity $v_y$.
- Label each vector with the numerical value of the velocity component at that point.
- The length of each vector should indicate its magnitude, using the length of the 10 m/s vector at $t = 0$ s as a reference.

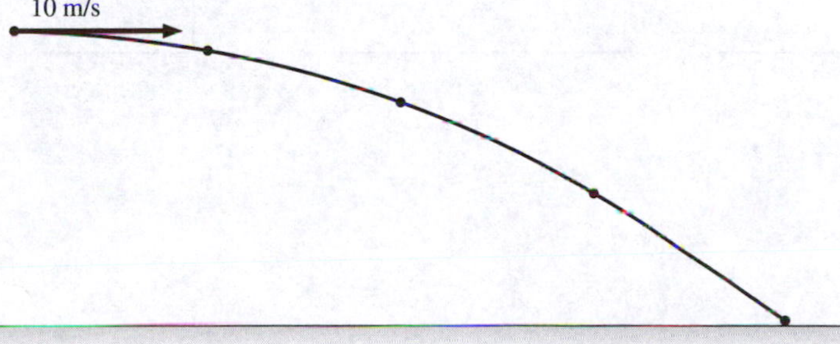

## 3.7 Projectile Motion: Solving Problems

38. a. A cart rolling at constant velocity fires a ball straight up. When the ball comes down, will it land in front of the launching tube, behind the launching tube, or directly in it? Explain.

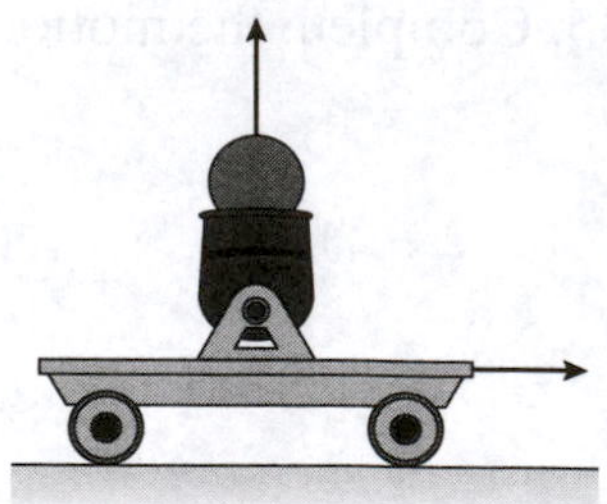

b. Will your answer change if the cart is accelerating in the forward direction? If so, how?

39. Four balls are simultaneously launched with the same speed from the same height $h$ above the ground. At the same instant, ball 5 is released from rest at the same height. Rank in order, from shortest to longest, the amount of time it takes each of these balls to hit the ground. (Some may be simultaneous.)

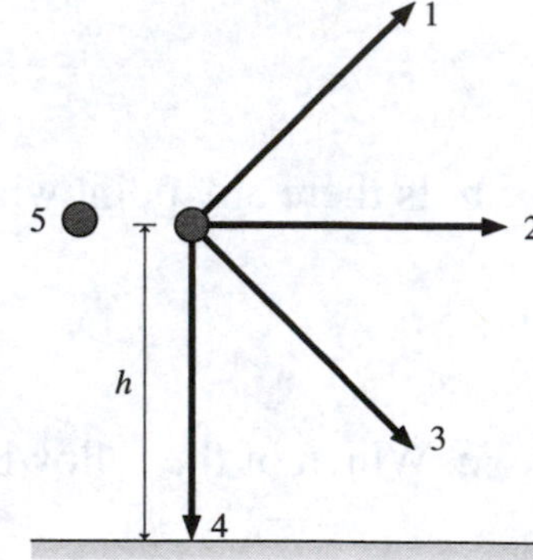


Order:

Explanation:

40. Rank in order, from shortest to longest, the amount of time it takes each of these projectiles to hit the ground. (Some may be simultaneous.)

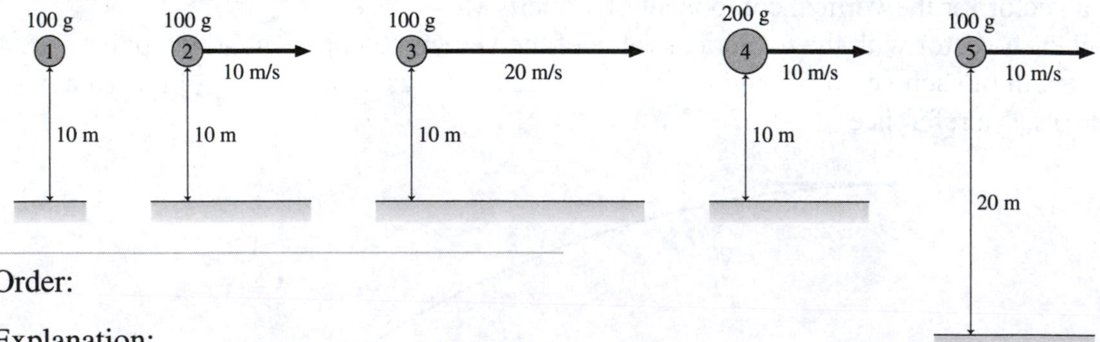


Order:

Explanation:

## 3.8 Motion in Two Dimensions: Circular Motion

41. The dots of a motion diagram are shown below for an object in uniform circular motion. Carefully complete the diagram.
    - Draw and label the velocity vectors $\vec{v}$. Use a **black** pen or pencil.
    - Draw and label the acceleration vectors $\vec{a}$. Use a **red** pen or pencil.

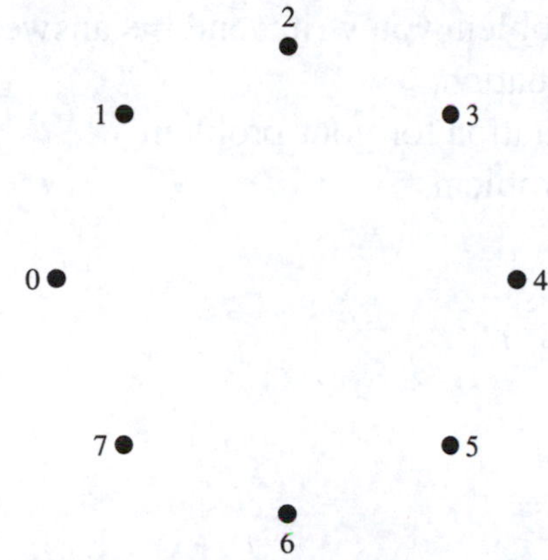


42. An object travels in a circle of radius $r$ at constant speed $v$.
    a. By what factor does the object's acceleration change if its speed is doubled and the radius is unchanged?

    b. By what factor does the acceleration change if the radius of the circle is doubled and its speed is unchanged?

43. The figure is a bird's-eye view of a car traveling at a steady 20 mph. At each of the three dots, labeled A, B, and C, either (a) write $\vec{a} = \vec{0}$ if the car is not accelerating or (b) draw and label a vector, with its tail at the dot, to show the car's acceleration.

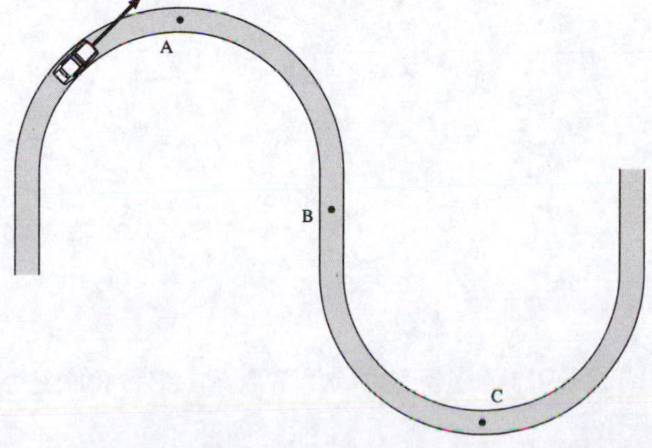

## You Write the Problem!

**Exercises 44–46:** You are given the equation or equations used to solve a problem. For each of these:

a. Write a realistic physics problem for which this is the correct equation. Look at worked examples and end-of-chapter problems in the textbook to see what realistic physics problems are like. Be sure that the problem you write, and the answer you ask for, is consistent with the information given in the equation.
b. Draw the pictorial representation for your problem.
c. Finish the solution of the problem.

44. $x_1 = 0\text{ m} + (30\text{ m/s})t_1$

$$0\text{ m} = 300\text{ m} - \tfrac{1}{2}(9.8\text{ m/s}^2)t_1^2$$

45. $x_1 = 0\text{ m} + (50\cos 30°\text{ m/s})t_1$

$$0\text{ m} = 0\text{ m} + (50\sin 30°\text{ m/s})t_1 - \tfrac{1}{2}(9.8\text{ m/s}^2)t_1^2$$

46. $2.5\text{ m/s}^2 = \dfrac{v^2}{10\text{ m}}$

# 4 Forces and Newton's Laws of Motion

## 4.1 Motion and Force

1. Using the particle model, draw the force a person exerts on a table when (a) pulling it to the right across a level floor with a force of magnitude $F$, (b) pulling it to the left across a level floor with force $2F$, and (c) *pushing* it to the right across a level floor with force $F$.

   a. Table pulled right with force $F$

   b. Table pulled left with force $2F$

   c. Table pushed right with force $F$

2. Two or more forces are shown on the objects below. Draw and label the net force $\vec{F}_{net}$.

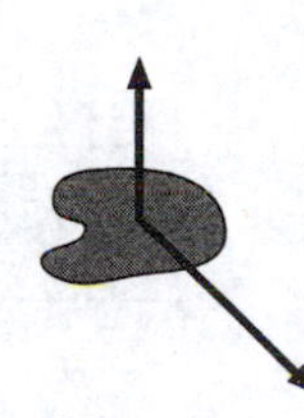

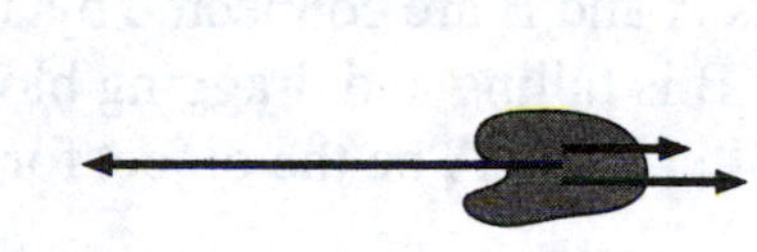

3. Two or more forces are shown on the objects below. Draw and label the net force $\vec{F}_{net}$.

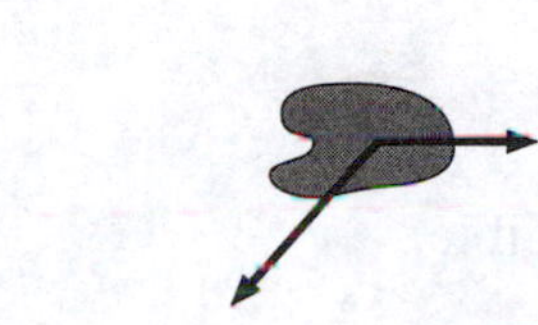

## 4.2 A Short Catalog of Forces

## 4.3 Identifying Forces

**Exercises 4–8:** Follow the six-step procedure of Tactics Box 4.2 to identify and name all the forces acting on the object.

4. An elevator suspended by a cable is descending at constant velocity.

5. A compressed spring is pushing a block across a rough horizontal table.

6. A brick is falling from the roof of a three-story building.

7. Blocks A and B are connected by a string passing over a pulley. Block B is falling and dragging block A across a frictionless table. Let block A be the object for analysis.

A

B

8. A rocket is launched at a 30° angle. Air resistance is not negligible.

## 4.4 What Do Forces Do?

9. The figure shows an acceleration-versus-force graph for an object of mass $m$. Data have been plotted as individual points, and a line has been drawn through the points.

   Draw and label, directly on the figure, the acceleration-versus-force graphs for objects of mass

   a. $2m$ b. $0.5m$

   Use triangles ▲ to show four points for the object of mass $2m$, then draw a line through the points. Use squares ■ for the object of mass $0.5m$.

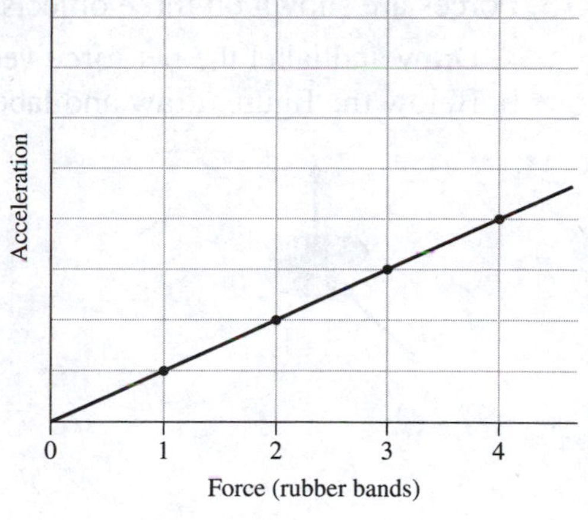


10. The quantity $y$ is inversely proportional to $x$, and $y = 4$ when $x = 9$.

    a. Write an equation to represent this inverse relationship for all $y$ and $x$.

    b. Find $y$ if $x = 12$. ________ c. Find $x$ if $y = 36$. ________

    d. Compare your equation in part a to the equation from your text relating $a$ and $m$, $a = \dfrac{F}{m}$.

    Which quantity assumes the role of $x$? ________

    Which quantity assumes the role of $y$? ________

    What is the constant of proportionality relating $a$ and $m$?

11. The quantity $y$ is inversely proportional to $x$. For one value of $x$, $y = 12$.

    a. What is the value of $y$ if $x$ is doubled? ________

    b. What is the value of $y$ if the original value of $x$ is halved? ________

12. A steady force applied to a 2.0 kg mass causes it to accelerate at 4.0 m/s$^2$. The same force is then applied to a 4.0 kg mass. Use ratio reasoning to find the acceleration of the 4.0 kg mass.

## 4.5 Newton's Second Law

13. Forces are shown on three objects. For each:
    a. Draw and label the net force vector. Do this right on the figure.
    b. Below the figure, draw and label the object's acceleration vector.

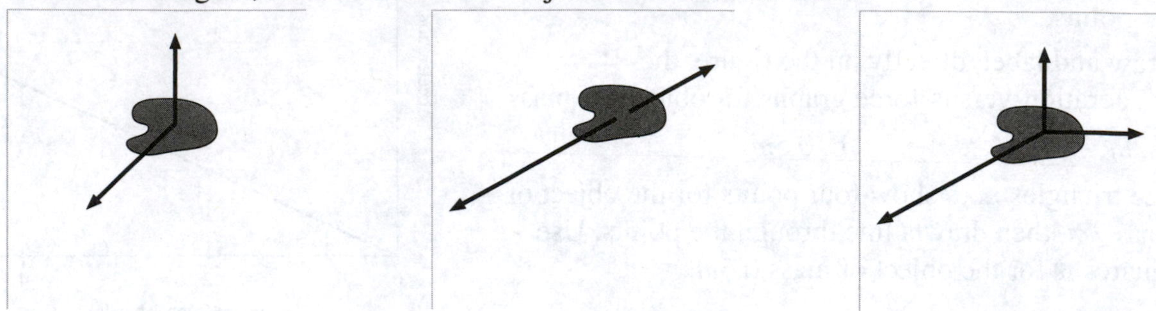

14. In the figures below, one force is missing. Use the given direction of acceleration to determine the missing force and draw it on the object. Do all work directly on the figure.

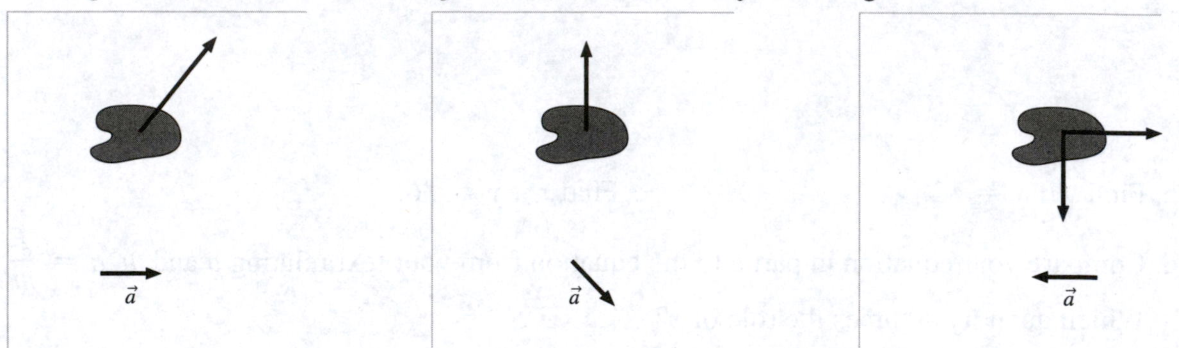


15. Below are two motion diagrams for a particle. Draw and label the net force vector at point 2.

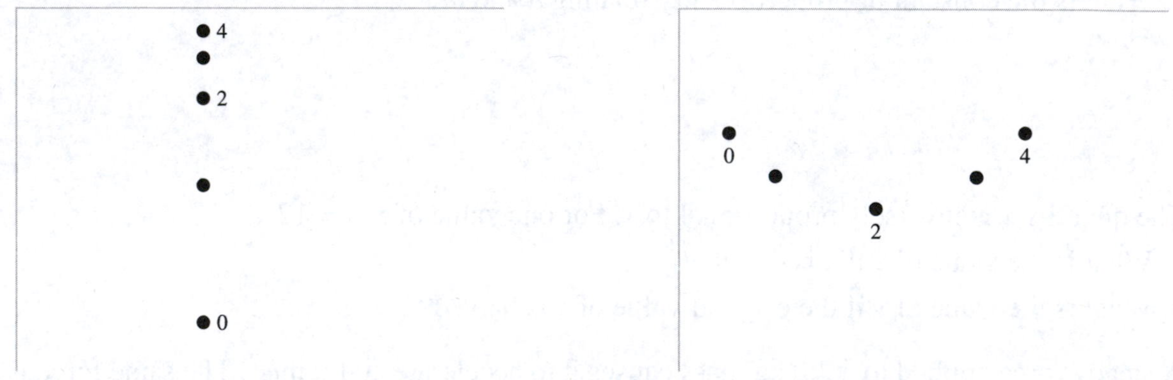


16. A constant force applied to an object causes the object to accelerate at 10 m/s$^2$. What will the acceleration of this object be if
    a. The force is doubled? ________ b. The mass is doubled? ________
    c. The force is doubled *and* the mass is doubled? ________
    d. The force is doubled *and* the mass is halved? ________

## 4.6 Free-Body Diagrams

**Exercises 17–22:**

- Draw a picture and identify the forces, following Tactics Box 4.2, then
- Draw a free-body diagram for the object, following each of the steps given in Tactics Box 4.3. Be sure to think carefully about the direction of $\vec{F}_{net}$.

**Note:** Draw individual force vectors with a **black** or **blue** pencil or pen. Draw the *net* force vector $\vec{F}_{net}$ with a **red** pencil or pen.

17. A heavy crate is being lowered straight down at a constant speed by a steel cable.

18. A boy is pushing a box across the floor at a steadily increasing speed. Let the box be the object for analysis.

19. A bicycle is speeding up down a hill. Friction is negligible, but air resistance is not.

20. You've slammed on your car brakes while going down a hill. The car is skidding to a halt.

21. You are going to toss a rock *straight up* into the air by placing it on the palm of your hand (you're not gripping it), then pushing your hand up very rapidly. You may want to toss an object into the air this way to help you think about the situation. The rock is the object of interest.

a. As you hold the rock at rest on your palm, before moving your hand.

b. As your hand is moving up but before the rock leaves your hand.

c. One-tenth of a second after the rock leaves your hand.

d. After the rock has reached its highest point and is now falling straight down.

22. Block B has just been released and is beginning to fall. Analyze block A.

A

B

## 4.7 Newton's Third Law

**Exercises 23–25:** Each of the following situations has two or more interacting objects. Draw a picture similar to Figure 4.30 in the textbook in which you

- Show the interacting objects, with a small gap separating them.
- Draw the force vectors of all action/reaction pairs.
- Label the force vectors, using a notation like $\vec{F}_{\text{A on B}}$ and $\vec{F}_{\text{B on A}}$.

23. A bat hits a ball. Draw your picture from the perspective of someone seeing the *end* of the bat at the moment it strikes the ball. The objects are the bat and the ball.

24. A boy pulls a wagon by its handle. Rolling friction is not negligible. The objects are the boy, the wagon, and the ground.

25. A crate is in the back of a truck as the truck accelerates forward. The crate does not slip. The objects are the truck, the crate, and the ground.

26. You find yourself in the middle of a frozen lake with a surface so slippery that you cannot walk. However, you happen to have several rocks in your pocket. The ice is extremely hard. It cannot be chipped, and the rocks slip on it just as much as your feet do. Can you think of a way to get to shore? Use pictures, forces, and Newton's laws to explain your reasoning.

27. How do basketball players jump straight up into the air? Your explanation should include pictures showing forces on the player and forces on the ground.

## You Write the Problem!

**Exercises 28–32:** You are given the free-body diagram of an object with one or more forces acting on it. For each of these:

a. Identify the direction of the object's acceleration $\vec{a}$. Draw and label the acceleration vector next to the free-body diagram. Or, if appropriate, write $\vec{a} = \vec{0}$.

b. Write a short description of a real object for which this is the correct free-body diagram. Use the worked examples in the textbook as models of what a description should be like.

28.

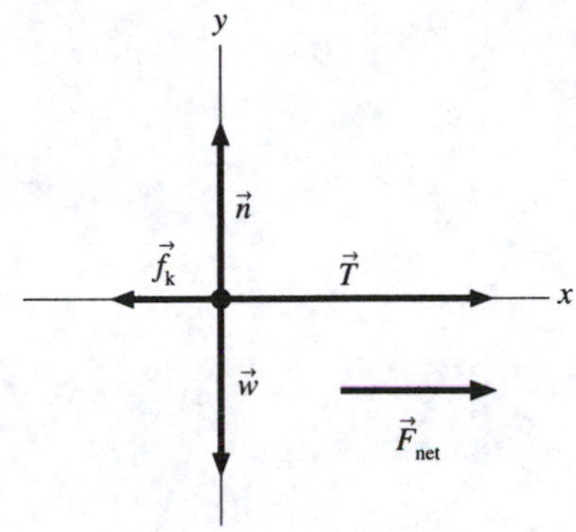


29.

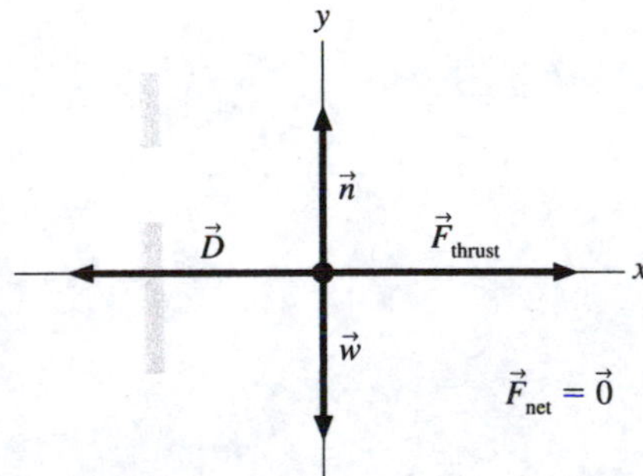


30.

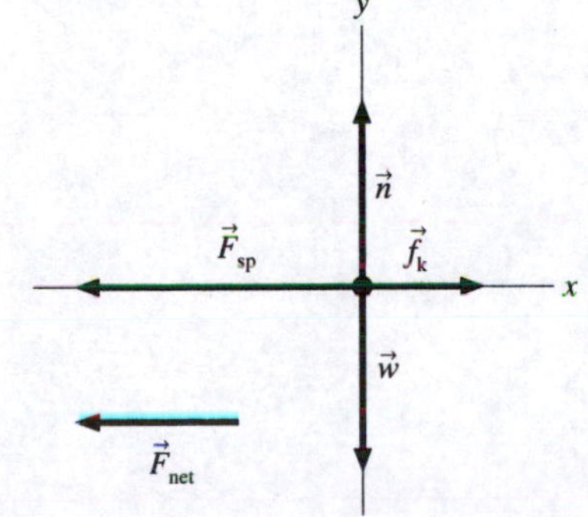

31.

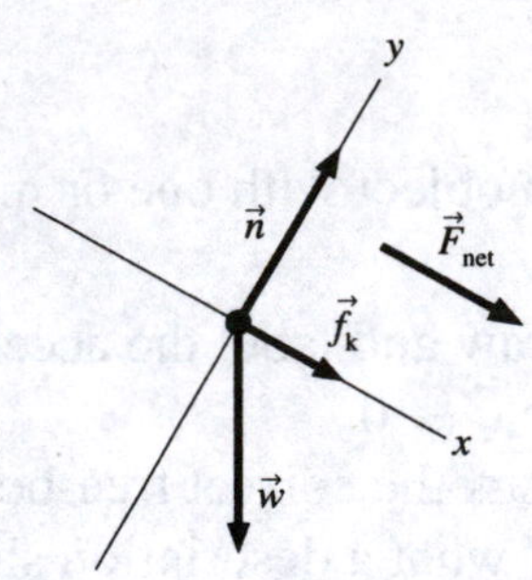


32.

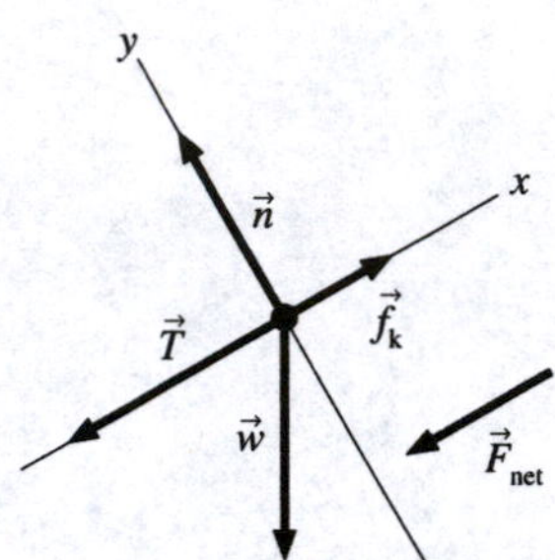

# 5 Applying Newton's Laws

## 5.1 Equilibrium

1. If an object is at rest, can you conclude that there are no forces acting on it? Explain.

2. If a force is exerted on an object, is it possible for that object to be moving with constant velocity? Explain.

3. A hollow tube forms three-quarters of a circle. It is lying flat on a table. A ball is shot through the tube at high speed. As the ball emerges from the other end, does it follow path A, path B, or path C? Explain your reasoning.

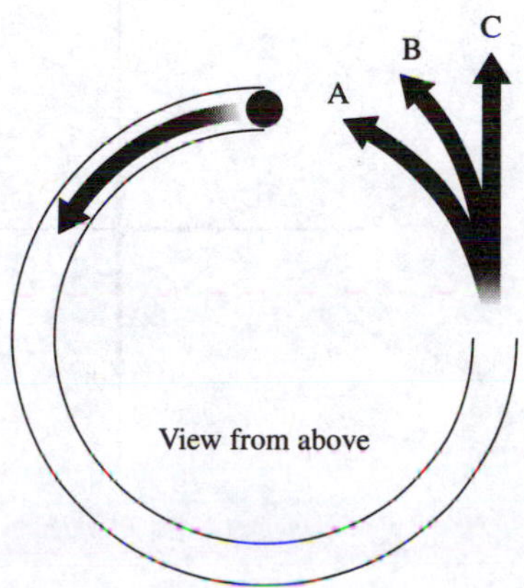

4. The vectors below show five forces that can be applied individually or in combinations to an object. Which forces or combinations of forces will cause the object to be in equilibrium?

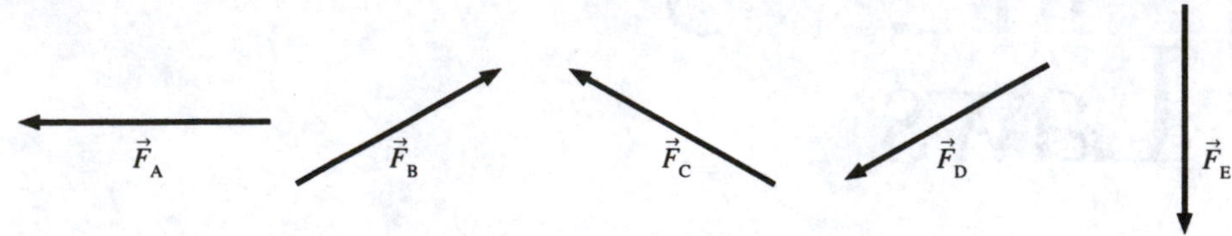

5. The free-body diagrams show a force or forces acting on an object. Draw and label one more force (one that is appropriate to the situation) that will cause the object to be in equilibrium.

a.

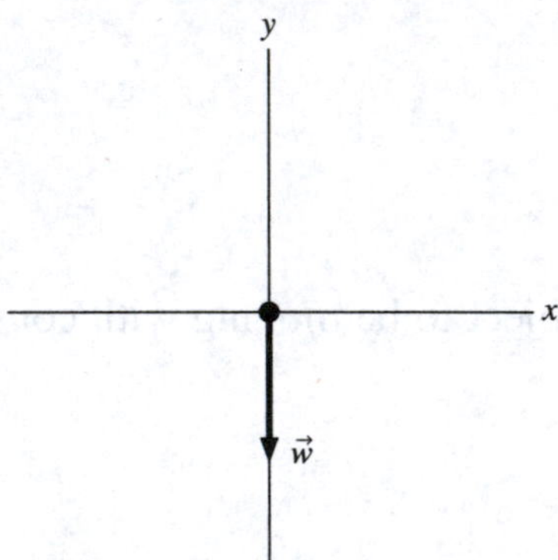


b.

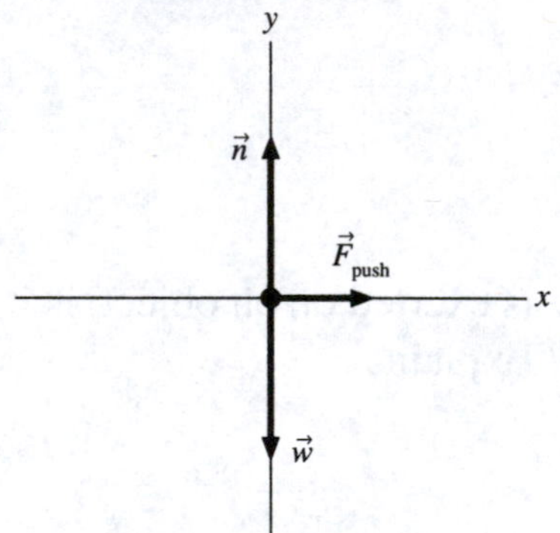


6. The free-body diagrams show a force or forces acting on an object. Draw and label one more force (one that is appropriate to the situation) that will cause the object to be in equilibrium.

a.

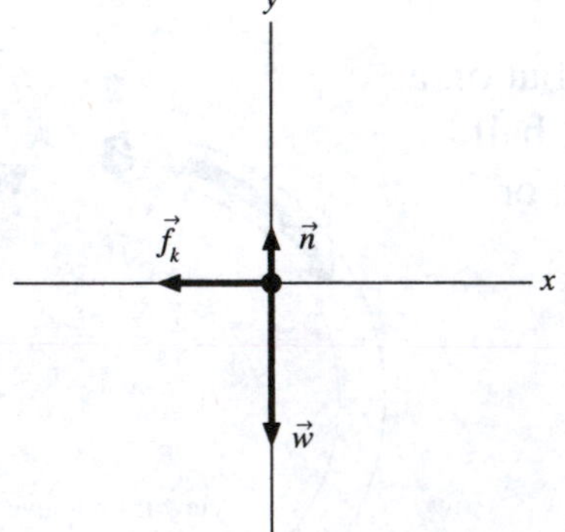


b.

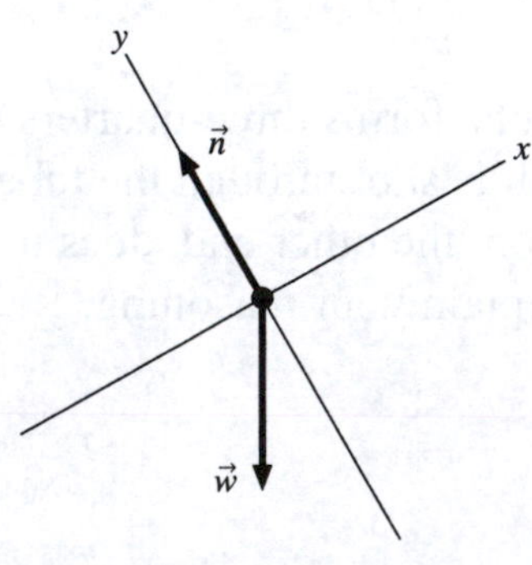

## 5.2 Dynamics and Newton's Second Law

7. a. An elevator travels *upward* at a constant speed. The elevator hangs by a single cable. Friction and air resistance are negligible. Is the tension in the cable greater than, less than, or equal to the weight of the elevator? Explain. Your explanation should include both a free-body diagram and reference to appropriate laws of physics.

b. The elevator travels *downward* and is slowing down. Is the tension in the cable greater than, less than, or equal to the weight of the elevator? Explain.

**Exercises 8–9:** The figures show free-body diagrams for an object of mass *m*. Write the *x*- and *y*-components of Newton's second law. Write your equations in terms of the *magnitudes* of the forces $F_1, F_2, \ldots$ and any *angles* defined in the diagram. One equation is shown in each question to illustrate the procedure.

8.

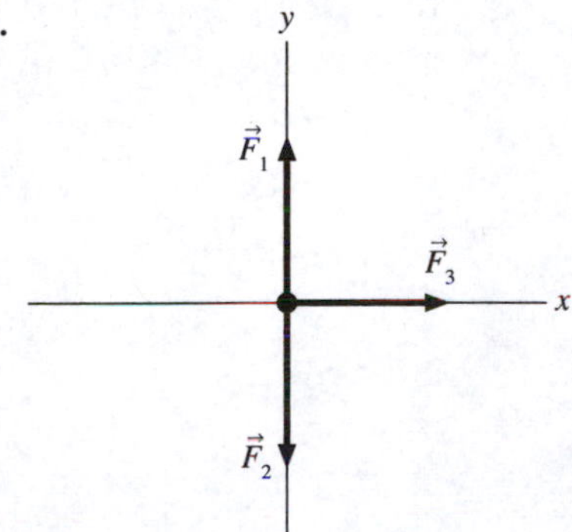


$ma_x =$

$ma_y = F_1 - F_2$

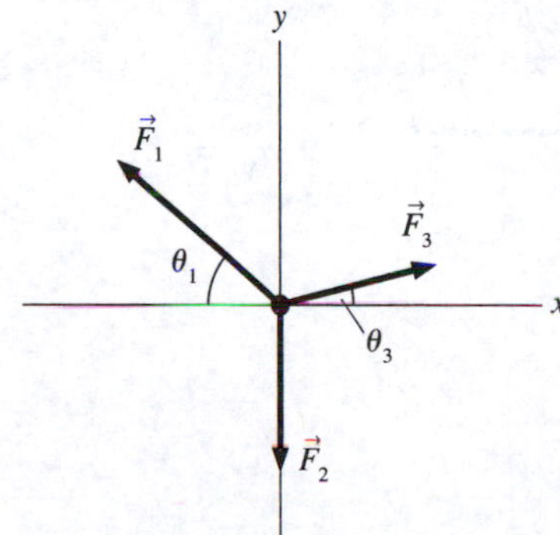


$ma_x =$

$ma_y =$

9.

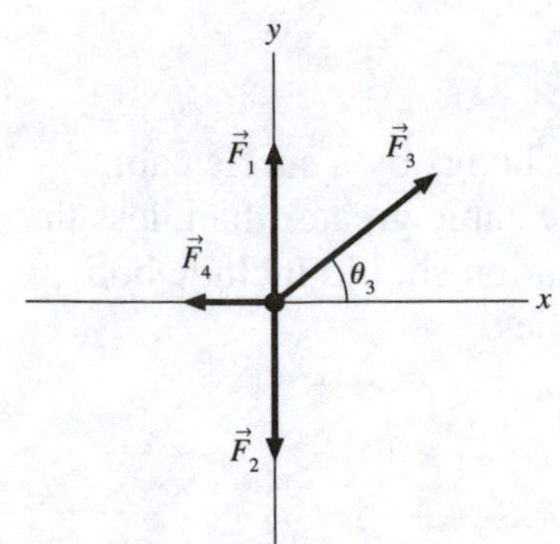


$ma_x = F_3 \cos\theta_3 - F_4$

$ma_y =$

$ma_x =$

$ma_y =$

**Exercises 10–12:** Two or more forces, shown on a free-body diagram, are exerted on a 2 kg object. The units of the grid are newtons. For each:

- Draw a vector arrow *on the grid*, starting at the origin, to show the net force $\vec{F}_{\text{net}}$.
- In the space to the right, determine the numerical values of the components $a_x$ and $a_y$.

10.

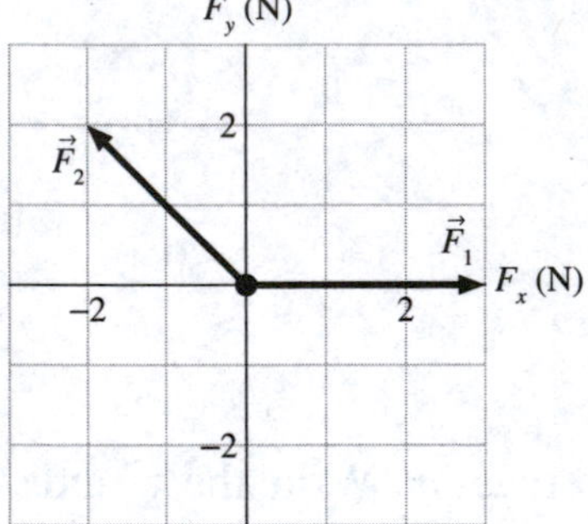


$a_x =$

$a_y =$

11.

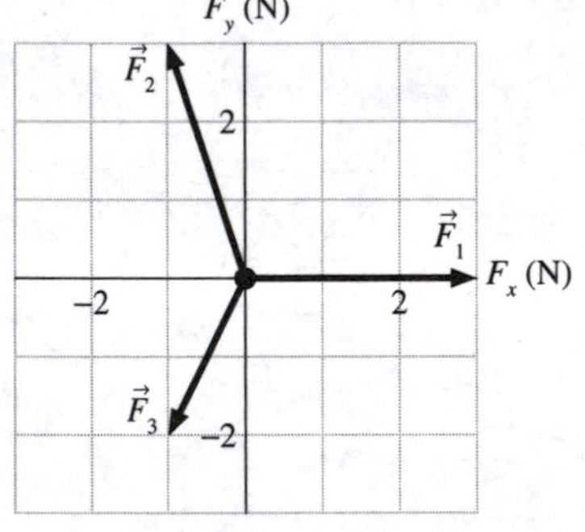


$a_x =$

$a_y =$

12.

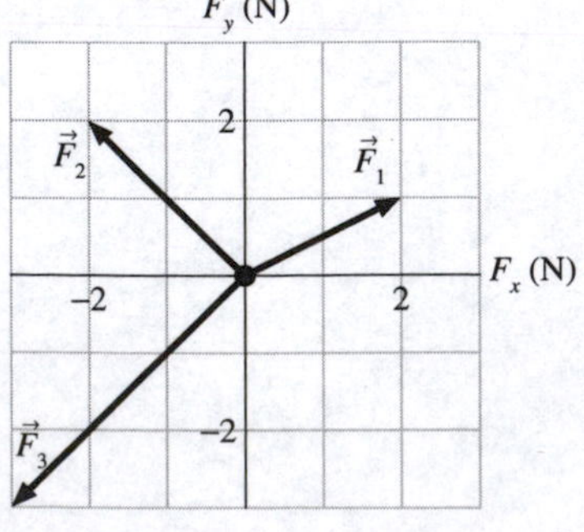


$a_x =$

$a_y =$

**Exercises 13–15:** Three forces $\vec{F}_1$, $\vec{F}_2$, and $\vec{F}_3$ cause a 1 kg object to accelerate with the acceleration given. Two of the forces are shown on the free-body diagrams below, but the third is missing. For each, draw and label *on the grid* the missing third force vector.

13. $a_x = 2 \text{ m/s}^2$
    $a_y = 0 \text{ m/s}^2$

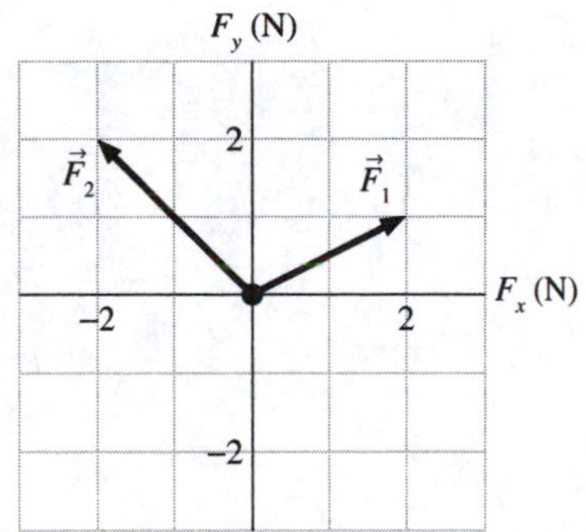


14. $a_x = 0 \text{ m/s}^2$
    $a_y = -3 \text{ m/s}^2$

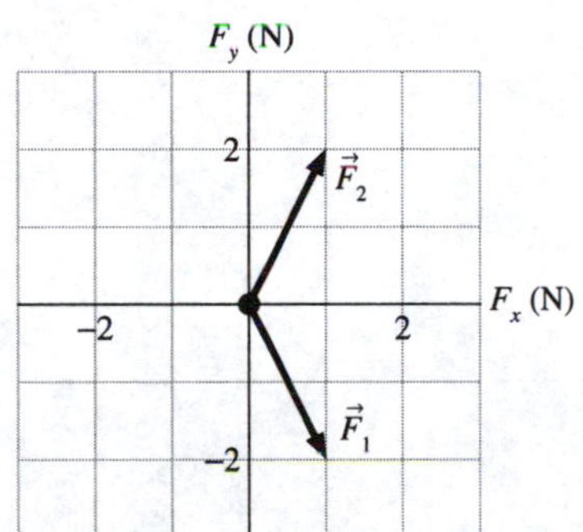


15. The object moves with constant velocity.

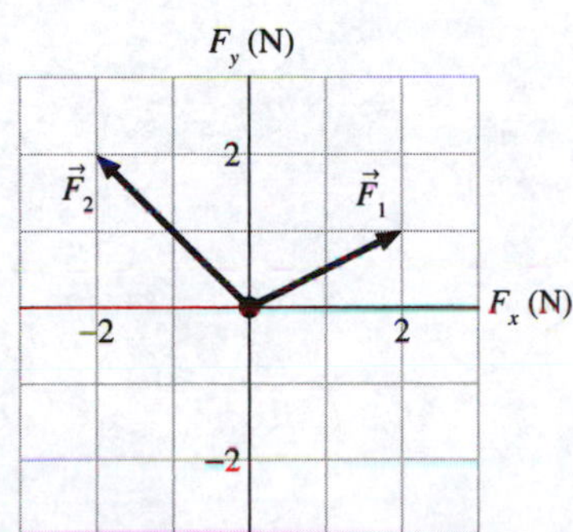

## 5.3 Mass and Weight

16. Suppose you have a jet-powered flying platform that can move straight up and down. For each of the following cases, is your apparent weight equal to, greater than, or less than your true weight? Explain.

    a. You are ascending and speeding up.

    b. You are descending and speeding up.

    c. You are ascending at a constant speed.

    d. You are ascending and slowing down.

    e. You are descending and slowing down.

17. The terms "vertical" and "horizontal" are frequently used in physics. Give *operational definitions* for these two terms. An operational definition defines a term by how it is measured or determined. Your definition should apply equally well in a laboratory or on a steep mountainside.

18. An astronaut orbiting the earth is handed two balls that are identical in outward appearance. However, one is hollow while the other is filled with lead. How might the astronaut determine which is which? Cutting them open is not allowed.

## 5.4 Normal Forces

19. Suppose you stand on a spring scale in six identical elevators. Each elevator moves as shown below. Let the reading of the scale in elevator $n$ be $S_n$. Rank in order, from largest to smallest, the six scale readings $S_1$ to $S_6$. Some may be equal. Give your answer in the form A > B = C > D.

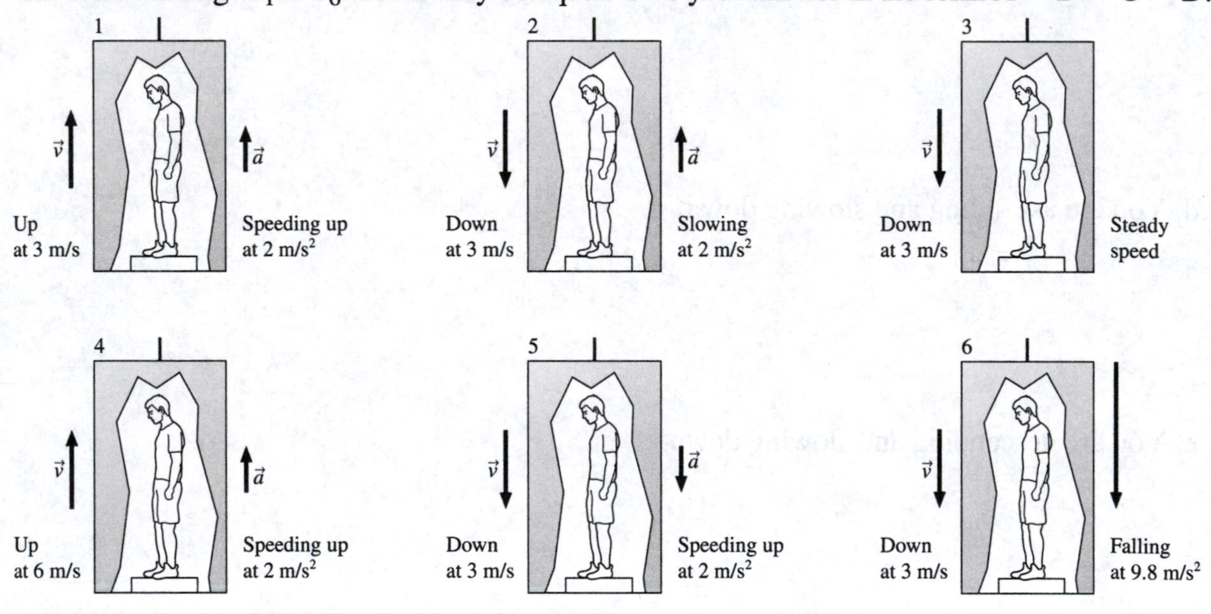


Order:

Explanation:

## 5.5 Friction

20. A block pushed along the floor with velocity $\vec{v}_0$ slides a distance $d$ after the pushing force is removed.
    a. If the mass of the block is doubled but the initial velocity is not changed, what is the distance the block slides before stopping? Explain.

    b. If the initial velocity of the block is doubled to $2\vec{v}_0$ but the mass is not changed, what is the distance the block slides before stopping? Explain.

21. Consider a box in the back of a pickup truck.
    a. If the truck accelerates slowly, the box moves with the truck without slipping. What force or forces act on the box to accelerate it? In what direction do those forces point?

    b. Draw a free-body diagram of the box.

    c. What happens to the box if the truck accelerates too rapidly? Explain why this happens, basing your explanation on the physical laws and models described in this chapter.

## 5.6 Drag

22. Five balls move through the air as shown. All five have the same size and shape. Rank in order, from largest to smallest, the magnitude of their accelerations $a_1$ to $a_5$. Some may be equal. Give your answer in the form A > B = C > D.

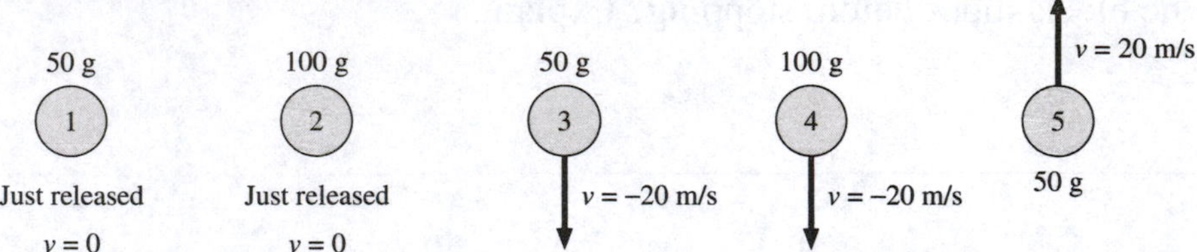


Order:

Explanation:

23. A 1 kg wood ball and a 5 kg metal ball have identical shapes and sizes. They are dropped simultaneously from a tall tower. Each ball experiences the *same* drag force because both have the same shape.

a. Draw free-body diagrams for the two balls as they fall in the presence of air resistance. Make sure your vectors all have the correct *relative* lengths.

b. Both balls have the same acceleration if they fall in a vacuum. The metal ball has a larger gravitational force, but it also has a larger mass and so $a = F/m$ is the same as for the wood ball. But the accelerations are not the same if the balls fall in air. Which now has the larger acceleration? Your explanation should refer explicitly to your free-body diagrams of part a and to Newton's second law.

c. Do the balls hit the ground simultaneously? If not, which hits first? Explain.

24. A small airplane of mass $m$ must take off from a primitive jungle airstrip that slopes upward at
PSS 5.2 a slight angle $\theta$. When the pilot pulls back on the throttle, the plane's engines exert a constant forward force $\vec{F}_{thrust}$. Rolling friction is not negligible on the dirt airstrip, and the coefficient of rolling resistance is $\mu_r$. If the plane's take-off speed is $v_{off}$, what minimum length must the airstrip have for the plane to get airborne?

a. Assume the plane takes off uphill to the right. Begin with a pictorial representation, as was described in Tactics Box 2.2. Establish a coordinate system with a tilted $x$-axis; show the plane at the beginning and end of the motion; define symbols for position, velocity, and time at these two points (six symbols all together); list known information; and state what you wish to find. $\vec{F}_{thrust}$, $m$, $\theta$, $\mu_r$, and $v_{off}$ are presumed known, although we have only symbols for them rather than numerical values, and three other quantities are zero.

b. Next, draw a force-identification diagram. Beside it, draw a free-body diagram. Your free-body diagram should use the same coordinate system you established in part a, and it should have 4 forces shown on it.

c. Write Newton's second law as two equations, one for the net force in the $x$-direction and one for the net force in the $y$-direction. Be careful finding the components of $\vec{w}$ (see Figure 5.11), and pay close attention to signs. Remember that symbols such as $w$ or $f_k$ represent the *magnitudes* of vectors; you have to supply appropriate signs to indicate which way the vectors point. The right side of these equations have $a_x$ and $a_y$. The motion is entirely along the $x$-axis, so what do you know about $a_y$? Use this information as you write the $y$-equation.

d. Now write the equation that characterizes the friction force on a rolling tire.

e. Combine your friction equation with the $y$-equation of Newton's second law to find an expression for the magnitude of the friction force.

f. Finally, substitute your answer to part e into the $x$-equation of Newton's second law, and then solve for $a_x$, the $x$-component of acceleration. Use $w = mg$ if you've not already done so.

g. With friction present, should the *magnitude* of the acceleration be larger or smaller than the acceleration of taking off on a frictionless runway? ________

h. Does your expression for acceleration agree with your answer to part g? ________
Explain how you can tell. If it doesn't, recheck your work.

i. The force analysis is done, but you still have to do the kinematics. This is a situation where we know about velocities, distance, and acceleration but nothing about the time involved. That should suggest the appropriate kinematics equation. Use your acceleration from part f in that kinematics equation, and solve for the unknown quantity you're seeking.

You've found a symbolic answer to the problem, one that you could now evaluate for a range of values of $\vec{F}_{\text{thrust}}$ or $\theta$ without having to go through the entire solution each time.

## You Write the Problem!

**Exercises 25–27:** You are given the dynamics equation that is used to solve a problem. For each of these:

a. Write a *realistic* physics problem for which this is the correct equation. Look at worked examples and end-of-chapter problems in the textbook to see what realistic physics problems are like. Be sure that the problem you write, and the answer you ask for, is consistent with the information given in the equation.
b. Draw the free-body diagram and pictorial representation for your problem.
c. Finish the solution of the problem.

25. $-0.80n = (1500\ \text{kg})a_y$

$n - (1500\ \text{kg})(9.8\ \text{m/s}^2) = 0$

26. $T - 0.20n - (20\ \text{kg})(9.8\ \text{m/s}^2)\sin 20° = (20\ \text{kg})(2.0\ \text{m/s}^2)$

$n - (20\ \text{kg})(9.8\ \text{m/s}^2)\cos 20° = 0$

27. $(100\ \text{N})\cos 30° - f_\text{k} = (20\ \text{kg})a_x$

$n + (100\ \text{N})\sin 30° - (20\ \text{kg})(9.8\ \text{m/s}^2) = 0$

$f_\text{k} = 0.20n$

## 5.7 Interacting Objects

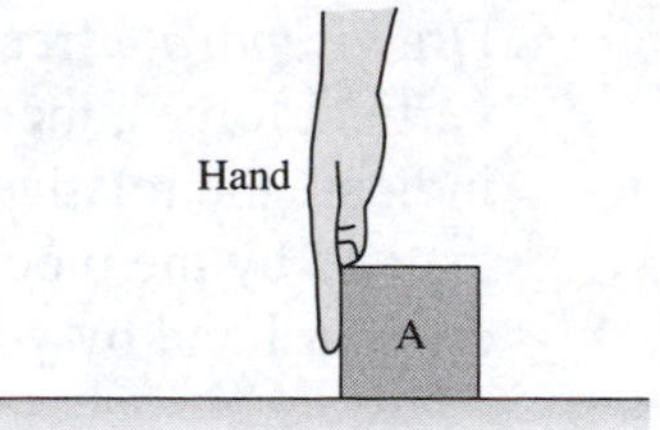


28. Block A is pushed across a horizontal surface at a *constant* speed by a hand that exerts force $\vec{F}_{\text{H on A}}$. The surface has friction.
    a. Draw two free-body diagrams, one for the hand and the other for the block. On these diagrams, show only the *horizontal* forces with lengths portraying the relative magnitudes of the forces. Label force vectors, using the form $\vec{F}_{\text{C on D}}$. On the hand diagram, show only $\vec{F}_{\text{H on A}}$. Don't include $\vec{F}_{\text{body on H}}$.

    b. Rank in order, from largest to smallest, the magnitudes of *all* of the horizontal forces you showed in part a. For example, if $F_{\text{C on D}}$ is the largest of three forces while $F_{\text{D on C}}$ and $F_{\text{D on E}}$ are smaller but equal, you can record this as $F_{\text{C on D}} > F_{\text{D on C}} = F_{\text{D on E}}$.

    Order:

    Explanation:

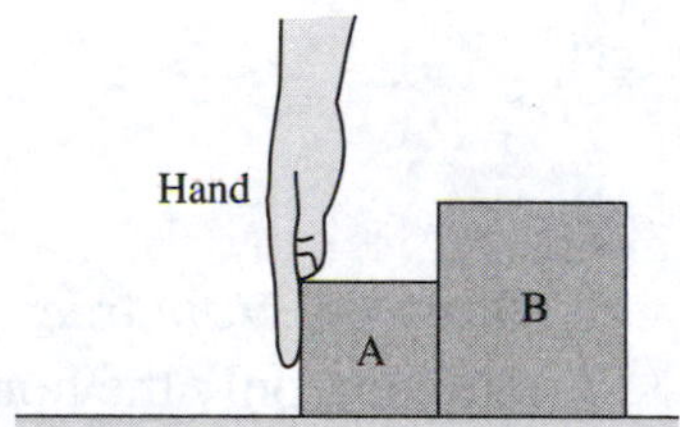


29. A second block B is placed in front of Block A of question 28. B is more massive than A: $m_B > m_A$. The blocks are speeding up.
    a. Consider a *frictionless* surface. Draw separate free-body diagrams for A, B, and the hand. Show only the horizontal forces. Label forces in the form $\vec{F}_{\text{C on D}}$.

    b. By applying Newton's second law to each block and the third law to each action/reaction pair, rank in order *all* of the horizontal forces, from largest to smallest.

    Order:

    Explanation:

30. Blocks A and B are held on the palm of your outstretched hand as you lift them straight up at *constant speed.* Assume $m_B > m_A$ and that $m_{hand} = 0$.

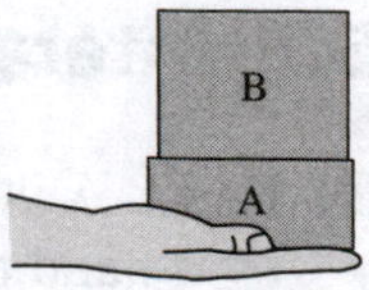

   a. Draw *separate* free-body diagrams for A, B, and your hand. Show *all* vertical forces, including the blocks' weights, making sure vector lengths indicate the relative sizes of the forces. For your hand, show only forces exerted by the blocks; neglect the weight of your hand or any forces exerted on your hand by your arm. Label forces in the form $\vec{F}_{C\text{ on }D}$.

   b. Rank in order, from largest to smallest, all of the vertical forces. Explain your reasoning.

31. A mosquito collides head-on with a car traveling 60 mph.

   a. How do you think the size of the force that the car exerts on the mosquito compares to the size of the force that the mosquito exerts on the car?

   b. Draw *separate* free-body diagrams of the car and the mosquito at the moment of collision, showing only the horizontal forces. Label forces in the form $\vec{F}_{C\text{ on }D}$.

   c. Does your answer to part b confirm your answer to part a? Explain why or why not.

## 5.8 Ropes and Pulleys

32. Blocks A and B are connected by a massless string over a massless, frictionless pulley. The blocks have just this instant been released from rest.

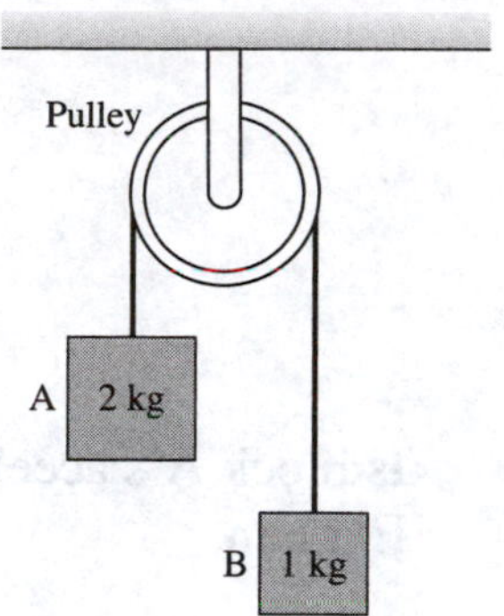


a. Will the blocks accelerate? If so, in which directions?

b. Draw a separate free-body diagram for each block. Be sure vector lengths indicate the relative sizes of the forces.

c. Rank in order, from largest to smallest, all of the vertical forces. Explain.

d. Consider the block that falls. Is the magnitude of its acceleration less than, greater than, or equal to $g$? Explain.

33. In case a, block A is accelerated across a frictionless table by a hanging 10 N weight (1.02 kg). In case b, the same block is accelerated by a steady 10 N tension in the string.

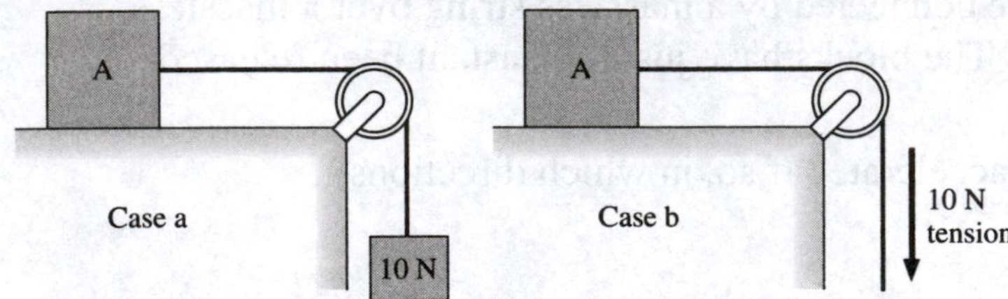


Is block A's acceleration in case b greater than, less than, or equal to its acceleration in case a? Explain.

**Exercises 34–35:** Draw separate free-body diagrams for blocks A and B. Indicate any pairs of forces having the same magnitude.

34.

35.

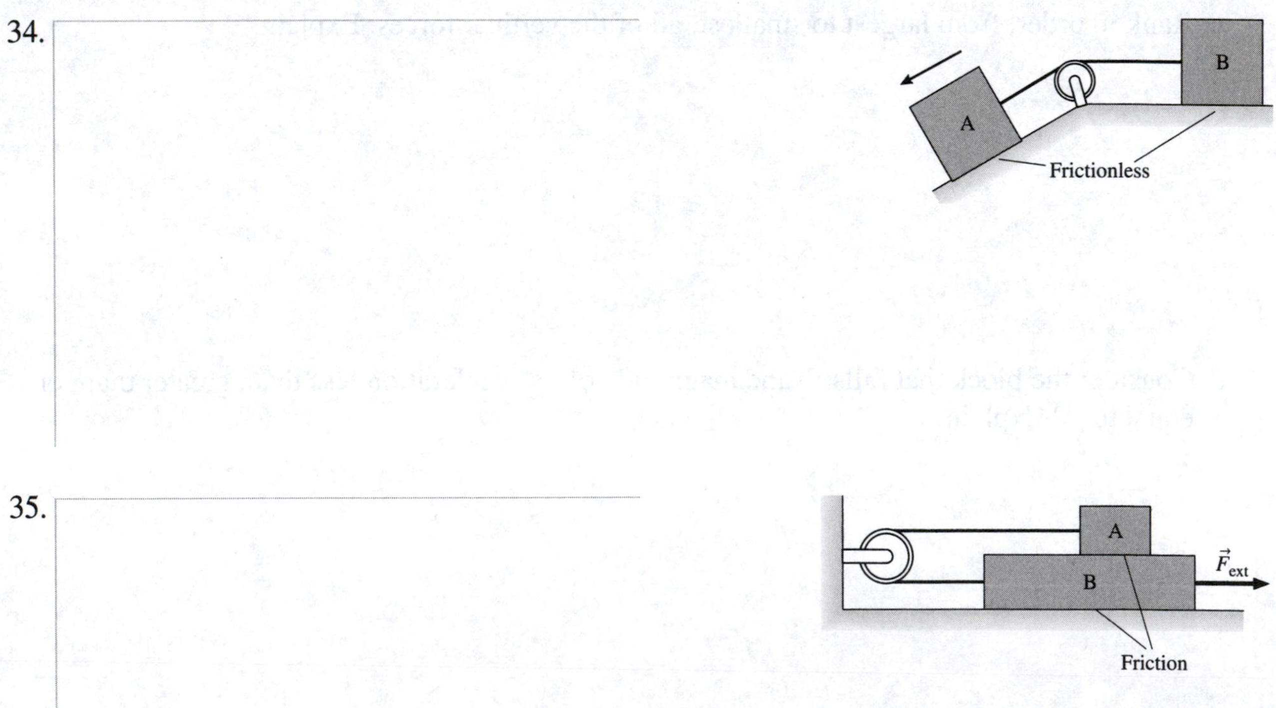

# 6 Circular Motion, Orbits, and Gravity

## 6.1 Uniform Circular Motion

1. a. The crankshaft in your car rotates at 3000 rpm. What is the frequency in revolutions per second?

   b. A record turntable rotates at 33.3 rpm. What is the rotation period in seconds?

2. The figure shows three points on a steadily rotating wheel.
   a. Draw the velocity vectors at each of the three points.
   b. Rank in order, from largest to smallest, the speeds $v_1$, $v_2$, and $v_3$ of these points.

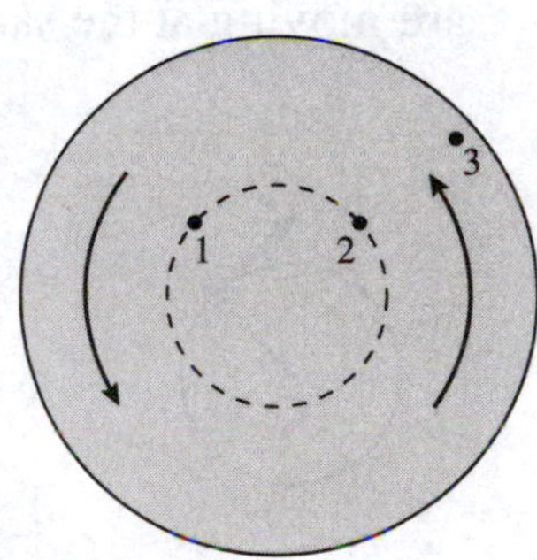


Order:

Explanation:

## 6.2 Dynamics of Uniform Circular Motion

3. The figure shows a *top view* of a plastic tube that is fixed on a horizontal table top. A marble is shot into the tube at A. Sketch the marble's trajectory after it leaves the tube at B. Explain.

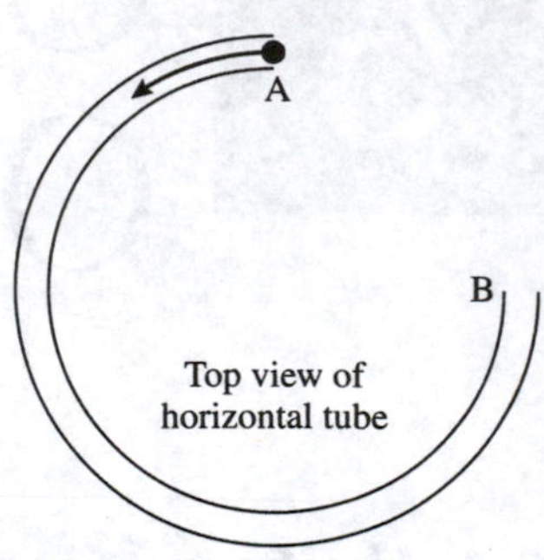


4. A ball swings in a *vertical* circle on a string. During one revolution, a very sharp knife is used to cut the string at the instant when the ball is at its lowest point. Sketch the subsequent trajectory of the ball until it hits the ground. Explain.

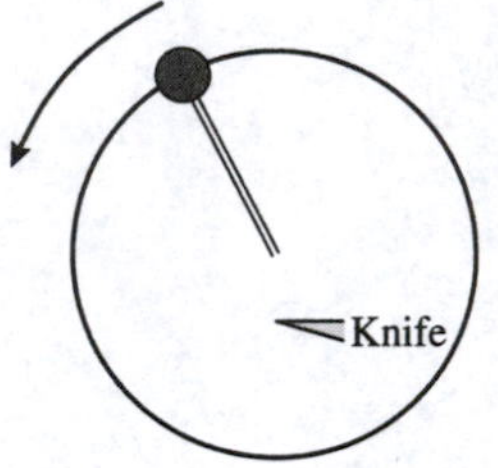


5. The figures are a bird's-eye view of particles moving in horizontal circles on a table top. All are moving at the same speed. Rank in order, from largest to smallest, the tensions $T_1$ to $T_4$.

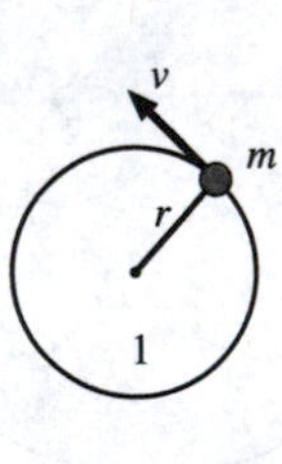


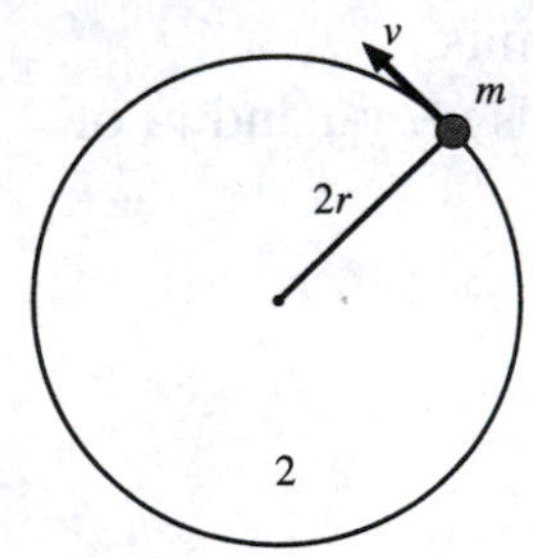


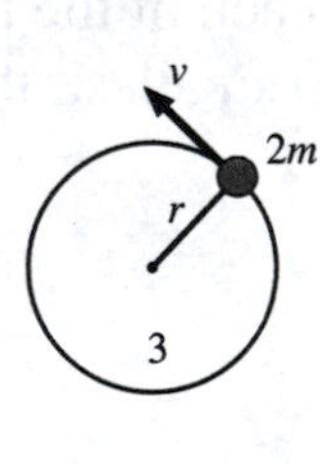


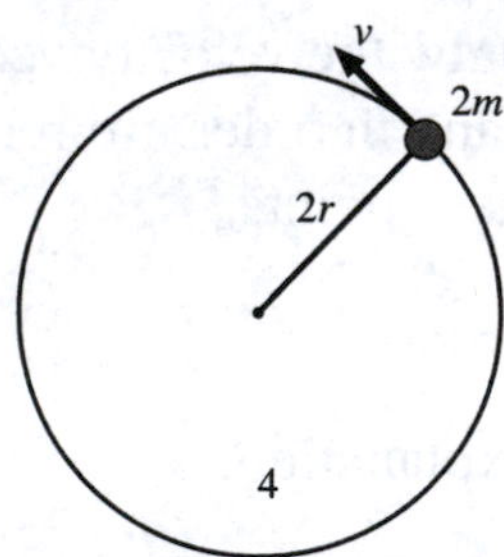


Order:

Explanation:

6. A ball rolls over the top of a circular hill. Rolling friction is negligible. Circle the letter of the ball's free-body diagram at the very top of the hill.

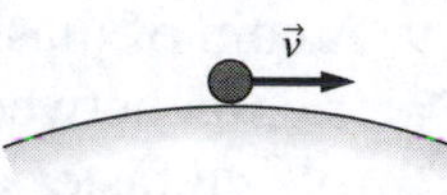

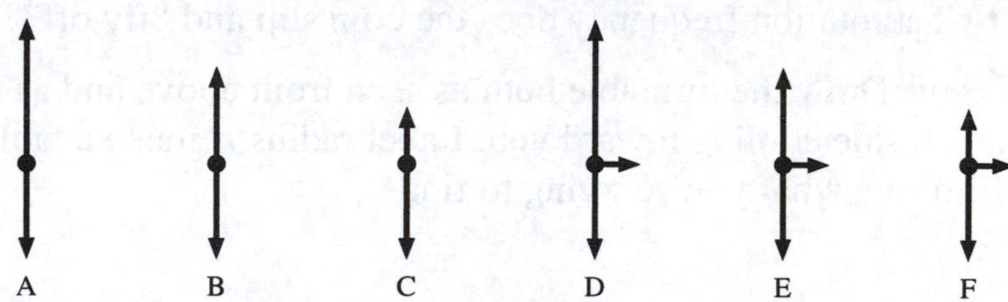


Explanation:

7. A ball on a string moves in a vertical circle. When the ball is at its lowest point, is the tension in the string greater than, less than, or equal to the ball's weight? Explain. (You should include a free-body diagram as part of your explanation.)

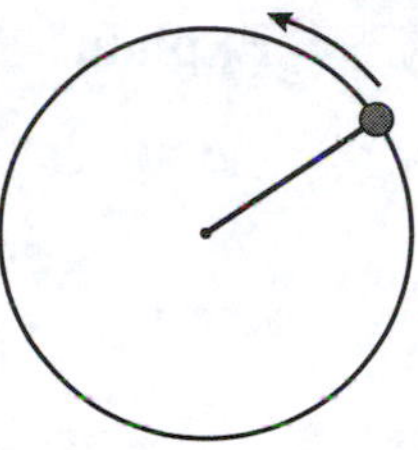

8. A marble rolls around the inside of a cone. Draw a free-body diagram of the marble when it is on the left side of the cone, coming toward you.

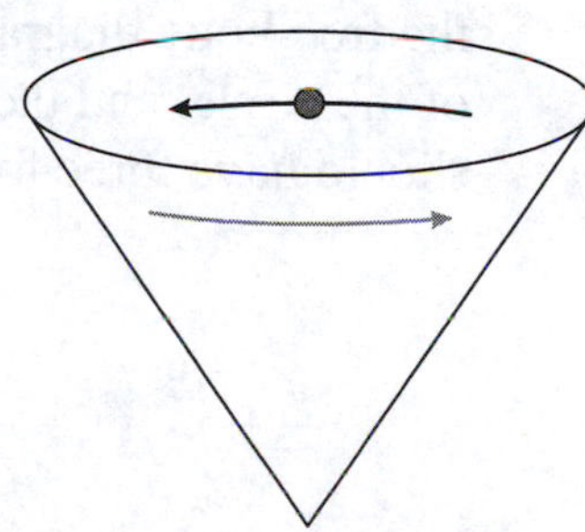

9. A coin of mass $m$ is placed at distance $r$ from the center of a turntable. The coefficient of static friction between the coin and the turntable is $\mu_s$. Starting from rest, the turntable is gradually rotated faster and faster. At what rotation frequency does the coin slip and "fly off"?

PSS 6.1

a. Begin with a visual overview. Draw the turntable both as seen from above and as an edge view with the coin on the left side coming toward you. Label radius $r$, make a table of known information, and indicate what you're trying to find.

b. What direction does $\vec{f}_s$ point? ______________________

Explain.

c. What condition describes the situation just as the coin starts to slip? Write this condition as a mathematical statement.

d. Now draw a free-body diagram of the coin. Following Problem Solving Strategy 6.1, draw the free-body diagram with the circle viewed edge on, the $x$-axis pointing toward the center of the circle, and the $y$-axis perpendicular to the plane of the circle. Your free-body diagram should have three forces on it.

e. Referring to Problem Solving Strategy 6.1, write Newton's second law for the $x$- and $y$-components of the forces. One sum should equal 0, the other $mv^2/r$.

f. The two equations of part e are valid for any speed up to the point of slipping. If you combine these with your statement of part c, you can solve for the speed $v_{max}$ at which the coin slips. Do so.

g. Finally, use the relationship between $v$ and $f$ to find the frequency at which the coin slips.

## 6.3 Apparent Forces in Circular Motion

10. The drawing is a partial motion diagram for a car rolling at constant speed over the top of a circular hill.
    a. Complete the motion diagram by adding the car's velocity vectors. Then use the velocity vectors to determine *and show* the car's acceleration $\vec{a}$ at the top of the hill.
    b. To the right, draw a free-body diagram of the car at the top of the hill. Next to the free-body diagram, indicate the direction of the net force on the car or, if appropriate, write $\vec{F}_{net} = \vec{0}$.

    c. Does the net force point of your free-body diagram point in the same direction you showed for the car's acceleration? ______
    If not, you may want to reconsider your work thus far because Newton's second law requires $\vec{F}_{net}$ and $\vec{a}$ to point the same way.
    d. Is there a maximum speed at which the car can travel over the top of the hill and not lose contact with the ground? If so, show how the free-body diagram would look at that speed. If not, why not?

11. A stunt plane does a series of vertical loop-the-loops. At what point in the circle does the pilot feel the heaviest? Explain. Include a free-body diagram with your explanation.

12. A roller-coaster car goes around the inside of a loop-the-loop. One of the following statements is true at the highest point in the loop, and one is true at the lowest point. Check the true statements.

| | Highest | Lowest |
|---|---|---|
| The car's apparent weight $w_{app}$ is always less than $w$ | | |
| The car's apparent weight $w_{app}$ is always equal to $w$ | | |
| The car's apparent weight $w_{app}$ is always greater than $w$ | | |
| $w_{app}$ could be less than, equal to, or greater than $w$ | | |

13. You can swing a ball on a string in a *vertical* circle if you swing it fast enough.
   a. Draw two free-body diagrams of the ball at the top of the circle. On the left, show the ball when it is going around the circle very fast. On the right, show the ball as it goes around the circle more slowly.

Very fast

Slower

   b. Suppose the ball has the smallest possible frequency that allows it to go all the way around the circle. What is the tension in the string when the ball is at the highest point? Explain.

## 6.4 Circular Orbits and Weightlessness

14. The earth has seasons because the axis of the earth's rotation is tilted 23° away from a line perpendicular to the plane of the earth's orbit. You can see this in the figure, which shows the edge of the earth's orbit around the sun. For both positions of the earth, draw a force vector to show the net force acting on the earth or, if appropriate, write $\vec{F} = \vec{0}$.

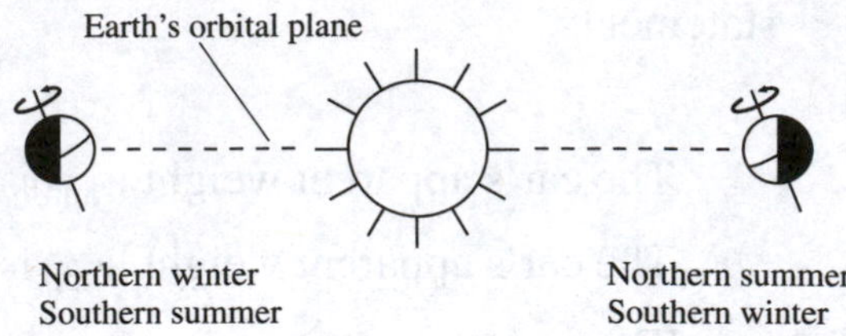


15. A small projectile is launched parallel to the ground at height $h = 1$ m with sufficient speed to orbit a completely smooth, airless planet. A bug rides in a small hole inside the projectile. Is the bug weightless? Explain.

16. It's been proposed that future space stations create "artificial gravity" by rotating around an axis. (The space station would have to be much larger than the present space station for this to be feasible.)

a. How would this work? Explain.

b. Would the artificial gravity be equally effective throughout the space station? If not, where in the space station would the residents want to live and work?

## 6.5 Newton's Law of Gravity

17. Is the earth's gravitational force on the sun larger than, smaller than, or equal to the sun's gravitational force on the earth? Explain.

18. Star A is twice as massive as star B.

   a. Draw gravitational force vectors on both stars. The length of each vector should be proportional to the size of the force.



   b. Is the acceleration of star A toward B larger than, smaller than, or equal to the acceleration of star B toward A? Explain.

19. The quantity $y$ is inversely proportional to the square of $x$, and $y = 4$ when $x = 5$.

   a. Write an equation to represent this inverse-square relationship for all $y$ and $x$.

   b. Find $y$ if $x = 2$. ________ c. Find $x$ if $y = 100$. ________

20. The quantity $y$ is inversely proportional to the square of $x$. For one value of $x$, $y = 12$.

   a. What is the value of $y$ if $x$ is doubled? ________

   b. What is the value of $y$ if the original value of $x$ is halved? ________

21. How far away from the earth does an orbiting spacecraft have to be in order for the astronauts inside to be "weightless"?

22. The acceleration due to gravity at the surface of Planet X is 20 $m/s^2$. The radius and the mass of Planet Z are twice those of Planet X. What is $g$ on Planet Z?

## 6.6 Gravity and Orbits

23. Planet X orbits the star Omega with a "year" that is 200 earth days long. Planet Y circles Omega at four times the distance of Planet X. How long is a year on Planet Y?

24. The mass of Jupiter is $M_{\text{Jupiter}} = 300M_{\text{earth}}$. Jupiter orbits around the sun with $T_{\text{Jupiter}} = 11.9$ years in an orbit with $r_{\text{Jupiter}} = 5.2r_{\text{earth}}$. Suppose the earth could be moved to the distance of Jupiter and placed in a circular orbit around the sun. The new period of the earth's orbit would be

a. 1 year.
b. 11.9 years.
c. Between 1 year and 11.9 years.
d. More than 11.9 years.
e. It could be anything, depending on the speed the earth is given.
f. It is impossible for a planet of earth's mass to orbit at the distance of Jupiter.

Circle the letter of the true statement. Then explain your choice.

25. The gravitational force of a star on orbiting planet 1 is $F_1$. Planet 2, which is twice as massive as Planet 1 and orbits at twice the distance from the star, experiences gravitational force $F_2$.

a. What is the ratio $F_2/F_1$?

b. Planet 1 orbits the star with period $T_1$ and Planet 2 with period $T_2$. What is the ratio $T_2/T_1$?

26. Satellite A orbits a planet with a speed of 10,000 m/s. Satellite B, orbiting at the same distance from the center of the planet, is twice as massive as Satellite A. What is the speed of Satellite B?

## You Write the Problem!

**Exercises 27–30:** You are given the equation that is used to solve a problem. For each of these:

a. Write a realistic physics problem for which this is the correct equation. Look at worked examples and end-of-chapter problems in the textbook to see what realistic physics problems are like. Be sure that the problem you write, and the answer you ask for, is consistent with the information given in the equation.
b. If appropriate, draw the free-body diagram and pictorial representation for your problem.
c. Finish the solution of the problem.

27. $60\text{ N} = \dfrac{(0.30\text{ kg})\,v^2}{0.50\text{ m}}$

28. $(1500\text{ kg})(9.8\text{ m/s}^2) - 11{,}760\text{ N} = \dfrac{(1500\text{ kg})\,v^2}{200\text{ m}}$

29. $\dfrac{(6.67 \times 10^{-11}\text{ N}\cdot\text{m}^2/\text{kg}^2)(1.90 \times 10^{27}\text{ kg})}{R^2} = \dfrac{(6.67 \times 10^{-11}\text{ N}\cdot\text{m}^2/\text{kg}^2)(5.98 \times 10^{24}\text{ kg})}{6.37 \times 10^{6}\text{ m}}$

30. $\dfrac{(6.67 \times 10^{-11}\text{ N}\cdot\text{m}^2/\text{kg}^2)(5.98 \times 10^{24}\text{ kg})(1000\text{ kg})}{r^2} = \dfrac{(1000\text{ kg})(1997\text{ m/s})^2}{r}$

# 7 Rotational Motion

## 7.1 Describing Circular and Rotational Motion

1. A particle undergoes uniform circular motion with constant angular velocity $\omega = +1.0$ rad/s, starting from point P.
   a. On the figure, draw a motion diagram showing the location of the particle every 1.0 s until the particle has moved through an angle of 5 rad. Draw velocity vectors **black** and acceleration vectors **red**. For this question, you can use 1 rad $\approx$ 60°.
   b. Below, graph the particle's angular position $\theta$ and angular velocity $\omega$ for the first 5 s of motion. Include an appropriate vertical scale on both graphs.

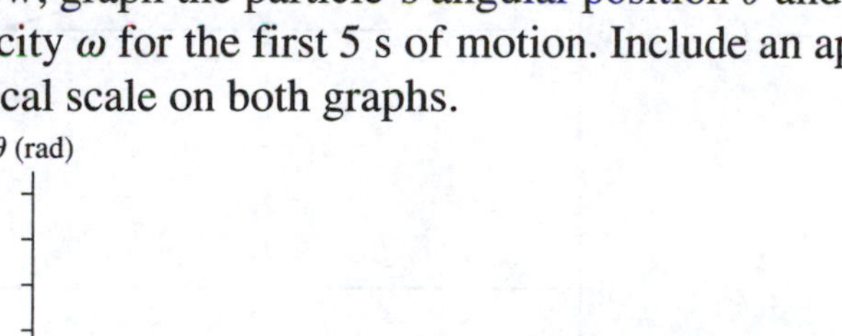


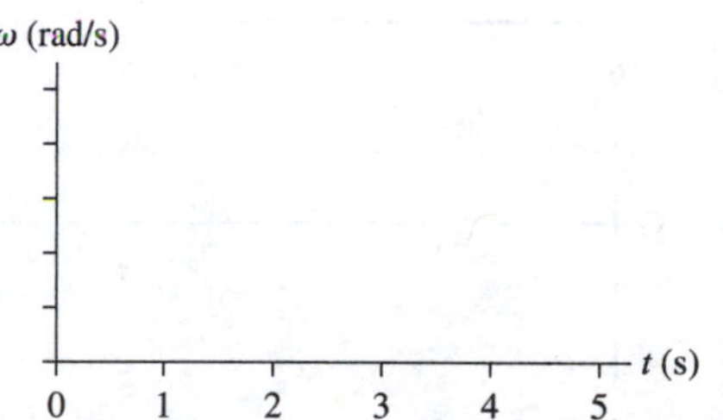

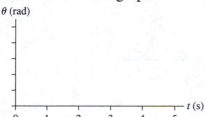

2. The figure shows three points on a steadily rotating wheel.
   a. Rank in order, from largest to smallest, the angular velocities $\omega_1$, $\omega_2$, and $\omega_3$ of these points.

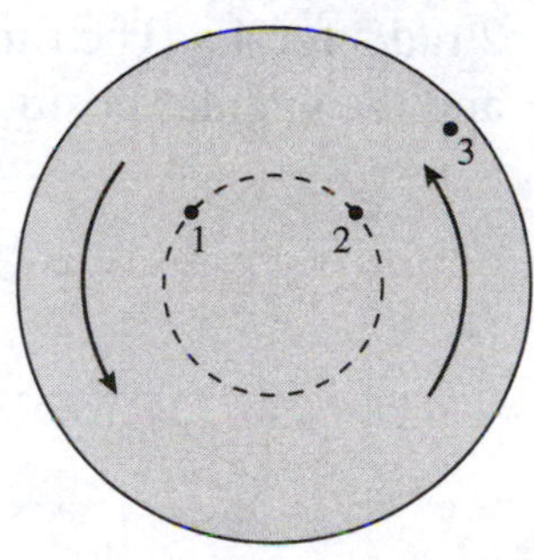


Order:

Explanation:

   b. Rank in order, from largest to smallest, the speeds $v_1$, $v_2$, and $v_3$ of these points.

Order:

Explanation:

3. Below are two angular position-versus-time graphs. For each, draw the corresponding angular velocity-versus-time graph directly below it.

a.

b.

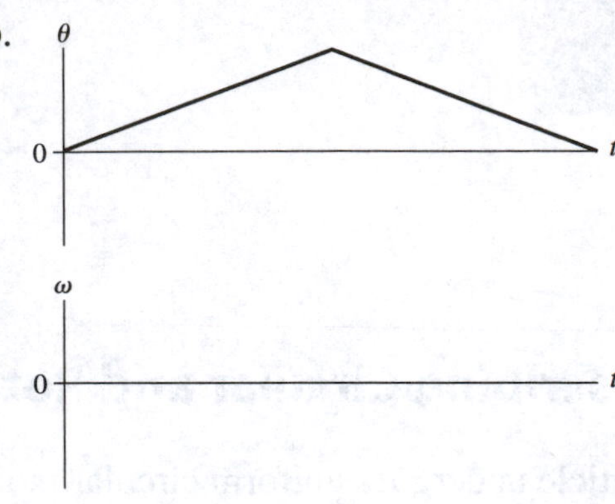


4. Below are two angular velocity-versus-time graphs. For each, draw the corresponding angular position-versus-time graph directly below it. Assume $\theta_0 = 0$ rad.

a.

b.

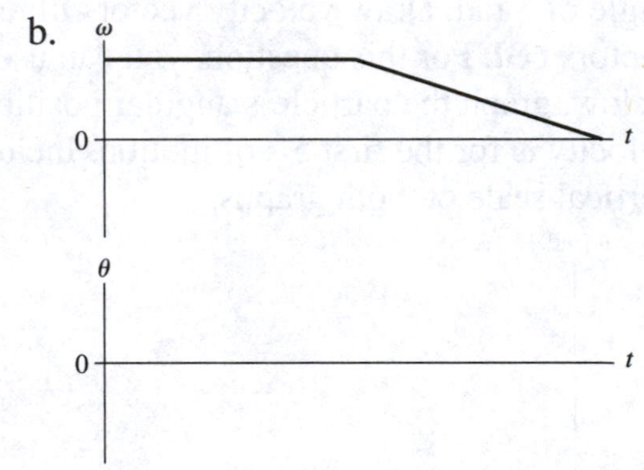


5. A particle in circular motion rotates clockwise at 4 rad/s for 2 s, and then counterclockwise at 2 rad/s for 4 s. The time required to change direction is negligible. Graph the angular velocity and the angular position, assuming $\theta_0 = 0$ rad.

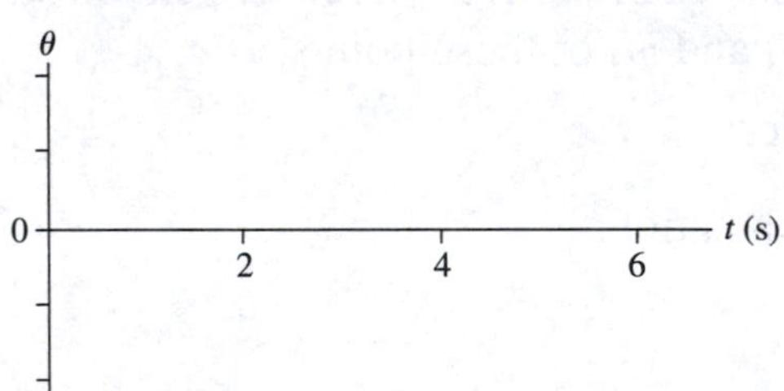


6. A particle moves in uniform circular motion with $a = 8\ \text{m/s}^2$. What is $a$ if

a. The radius is doubled without changing the angular velocity? ______

b. The radius is doubled without changing the particle's speed? ______

c. The angular velocity is doubled without changing the circle's radius? ______

## 7.2 The Rotation of a Rigid Body

7. The following figures show a rotating wheel. If we consider a counterclockwise rotation as positive (+) and a clockwise rotation as negative (−), determine the signs (+ or −) of $\omega$ and $\alpha$.

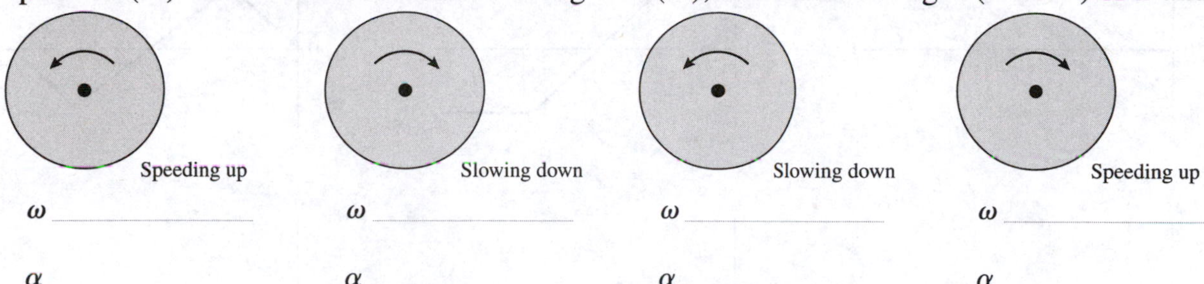


$\omega$ ________ $\omega$ ________ $\omega$ ________ $\omega$ ________

$\alpha$ ________ $\alpha$ ________ $\alpha$ ________ $\alpha$ ________

8. A ball is rolling back and forth inside a bowl. The figure shows the ball at extreme left edge of the ball's motion as it changes direction.

   a. At this point, is $\omega$ positive, negative, or zero? ________

   b. At this point, is $\alpha$ positive, negative, or zero? ________

9. The figures below show the centripetal acceleration vector $\vec{a}_c$ at four successive points on the trajectory of a particle moving in a counterclockwise circle.

   a. For each, draw the tangential acceleration vector $\vec{a}_t$ at points 2 and 3 or, if appropriate, write $\vec{a}_t = \vec{0}$.

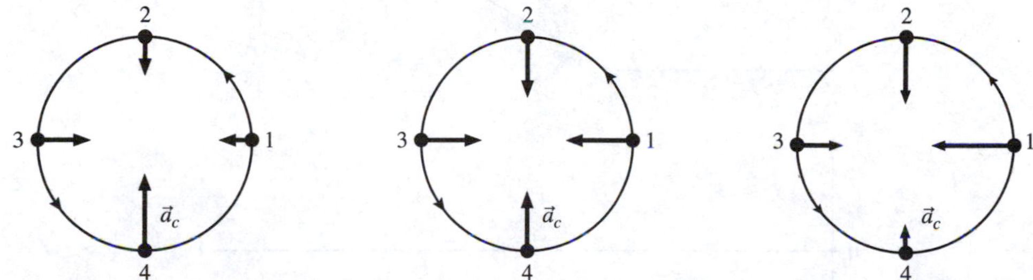


   b. If we consider a counterclockwise rotation as positive and a clockwise rotation as negative, determine if the particle's angular acceleration $\alpha$ is positive (+), negative (−), or zero (0).

$\alpha =$ ________ $\alpha =$ ________ $\alpha =$ ________

10. Below are three angular velocity-versus-time graphs. For each, draw the corresponding angular acceleration-versus-time graph.

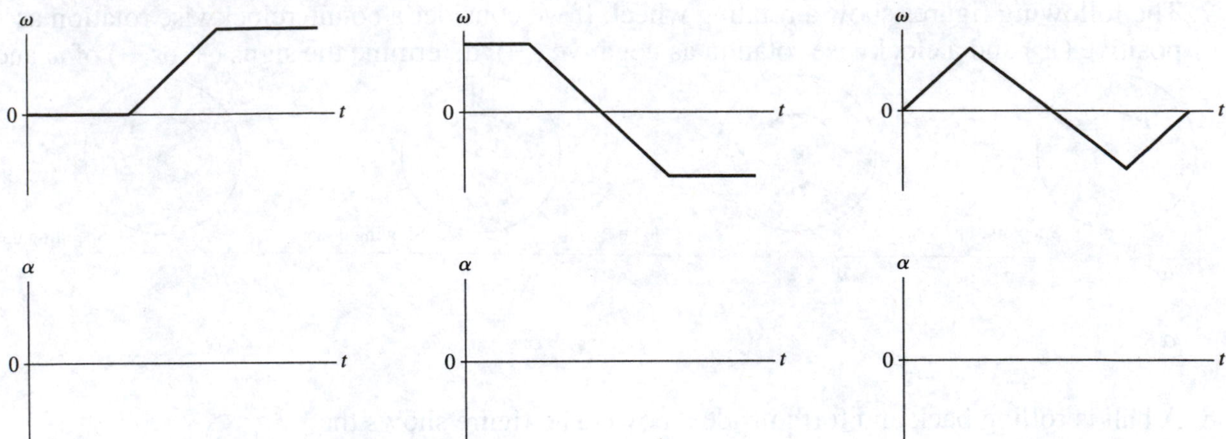


11. A wheel rolls to the left along a horizontal surface, down a ramp, and then continues along the lower horizontal surface. Draw graphs for the wheel's angular velocity $\omega$ and angular acceleration $\alpha$ as functions of time.

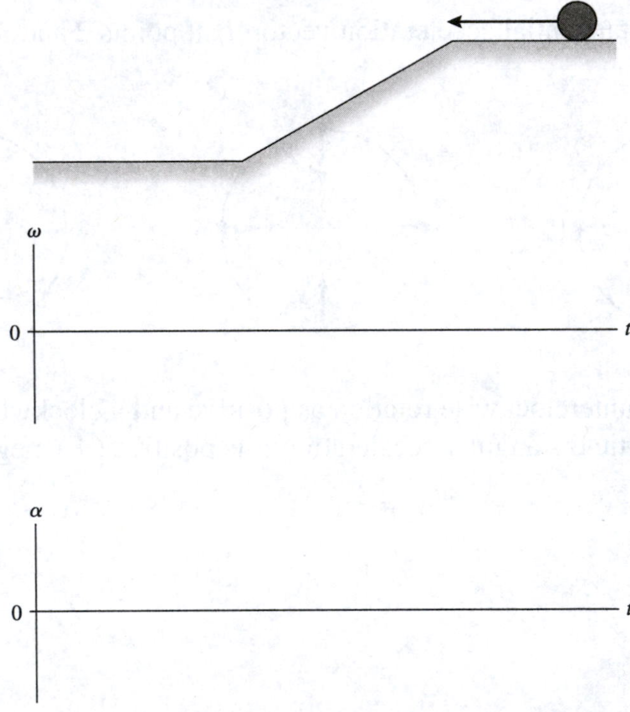

## 7.3 Torque

12. Five forces are applied to a door. For each, determine if the torque about the hinge is positive (+), negative (−), or zero (0).

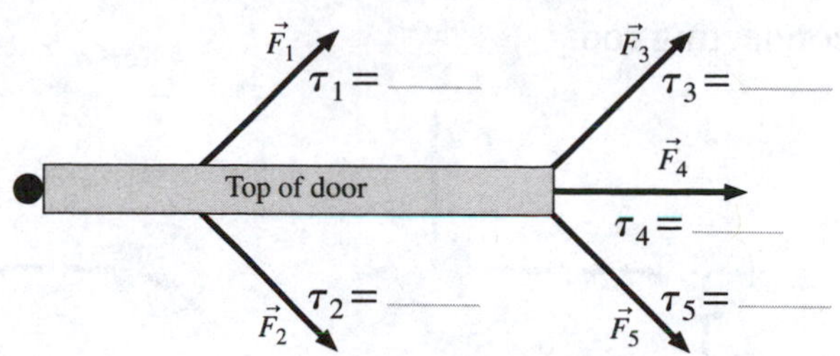


13. Six forces, each of magnitude either $F$ or $2F$, are applied to a door. Rank in order, from largest to smallest, the six torques $\tau_1$ to $\tau_6$ about the hinge.

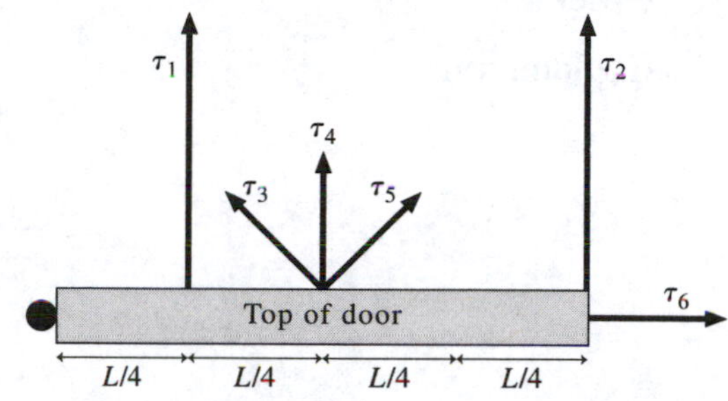


Order:

Explanation:

14. Four forces are applied to a rod that can pivot on an axle. For each force,
    a. Use a **black** pen or pencil to draw the line of action.
    b. Use a **red** pen or pencil to draw and label the moment arm, or state that $r_\perp = 0$.
    c. Determine if the torque about the axle is positive (+), negative (−), or zero (0).

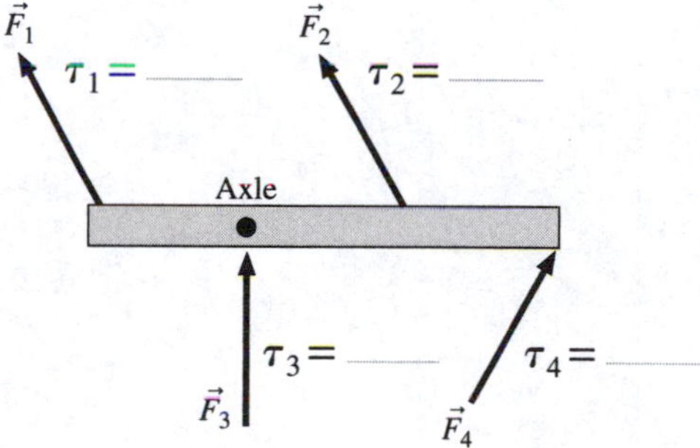


15. a. Draw a force vector at A whose torque about the axle is negative.
    b. Draw a force vector at B whose torque about the axle is zero.
    c. Draw a force vector at C whose torque about the axle is positive.

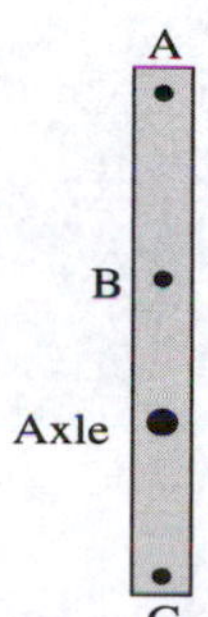

16. The dumbbells below are all the same size, and the forces all have the same magnitude. Rank in order, from largest to smallest, the torques $\tau_1$, $\tau_2$, and $\tau_3$ about the midpoint of each connecting rod.

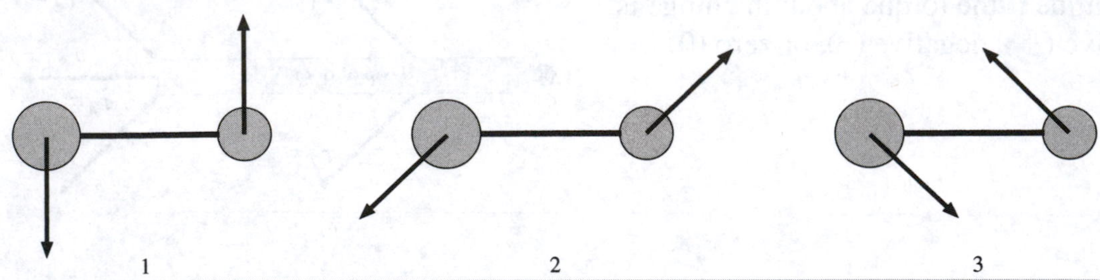


Order:

Explanation:

17. A bicycle is at rest on a smooth surface. A force is applied to the bottom pedal as shown. Does the bicycle roll forward (to the right), backward (to the left), or not at all? Explain.

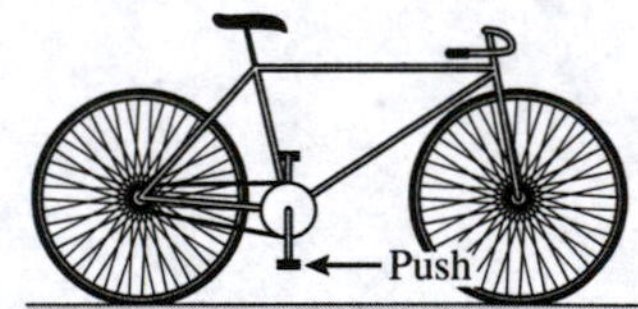

## 7.4 Gravitational Torque and the Center of Gravity

18. Two spheres are connected by a rigid but massless rod to form a dumbbell. The two spheres have equal mass. Mark the approximate location of the center of gravity with an ×. Explain your reasoning.

19. Two spheres are connected by a rigid but massless rod to form a dumbbell. The two spheres are made of the same material. Mark the approximate location of the center of gravity with an ×. Explain your reasoning.

20. Mark the center of gravity of this object with an ×. Explain.

21. a. Find the coordinates for and *mark* the center of gravity for the pair of masses shown, using the center of the 3.0 kg mass as the origin.

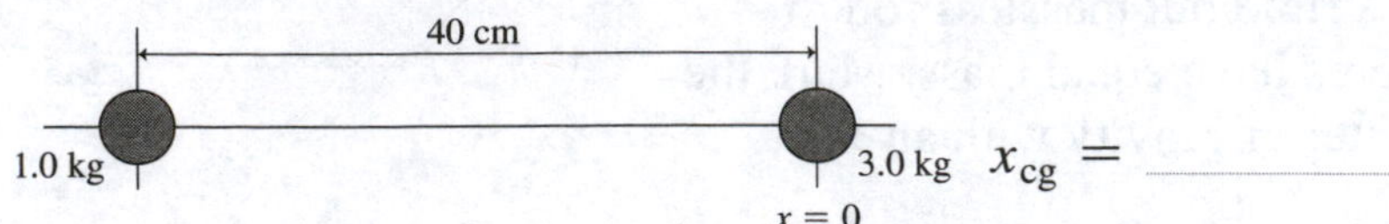


$x_{cg}$ = ________

b. Find the coordinates for and *mark* the center of gravity for the pair of masses shown, using the center of the 1.0 kg mass as the origin.

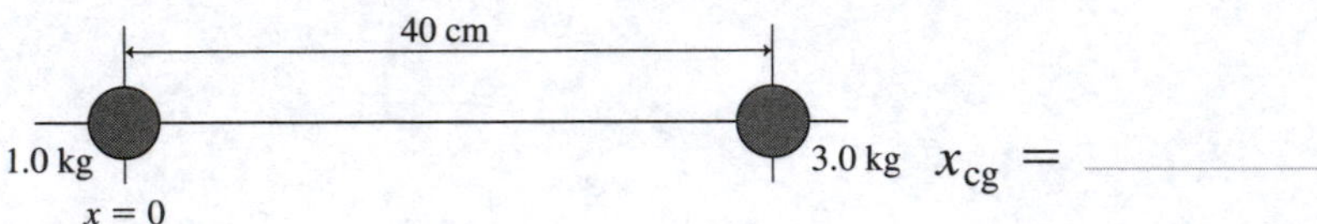


$x_{cg}$ = ________

c. How do the locations for the marks in parts a and b compare? How do the coordinates compare?

d. The 3.0 kg mass from parts a and b above is separated into two 1.5 kg pieces. One of these is moved 10 cm in the +*y*-direction. Find the coordinates for and *mark* the center of gravity of the new system using the center of the 1.0 kg mass as the origin.

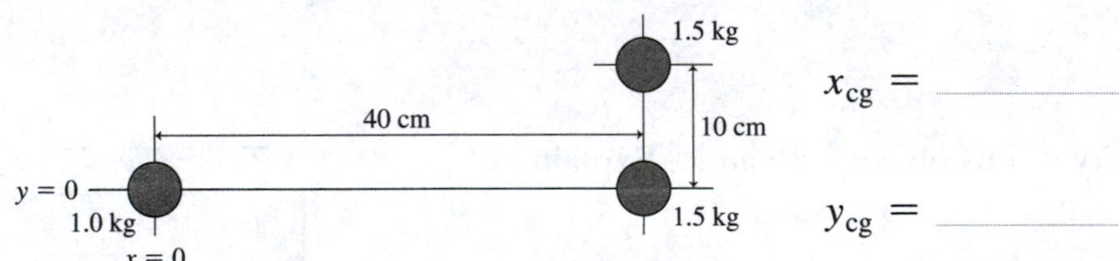


$x_{cg}$ = ________

$y_{cg}$ = ________

e. What effect did separating the 3.0 kg mass along the *y*-direction have on the *x*-component of the center of gravity of the system?

## 7.5 Rotational Dynamics and Moment of Inertia

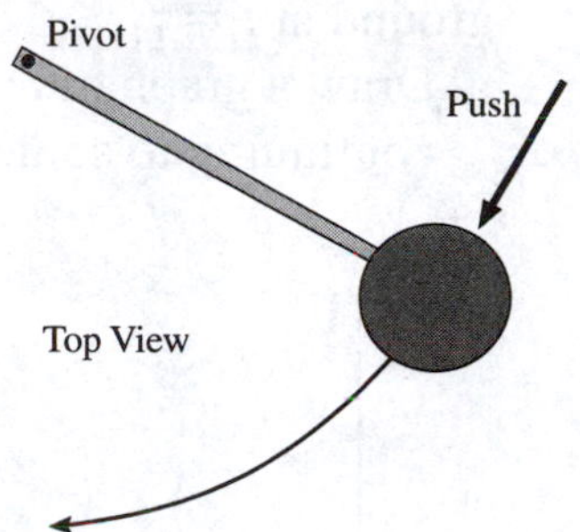


22. A student gives a quick push to a ball at the end of a massless, rigid rod, causing the ball to rotate clockwise in a *horizontal* circle. The rod's pivot is frictionless.
   a. As the student is pushing, is the torque about the pivot positive, negative, or zero? ______
   b. After the push has ended, does the ball's angular velocity
      i. Steadily increase?
      ii. Increase for awhile, then hold steady?
      iii. Hold steady?
      iv. Decrease for awhile, then hold steady?
      v. Steadily decrease?

      Explain the reason for your choice.

   c. Right after the push has ended, is the torque positive, negative, or zero? ______

23. The top graph shows the torque on a rotating wheel as a function of time. The wheel's moment of inertia is 10 kg · m$^2$. Draw graphs of $\alpha$-versus-$t$ and $\omega$-versus-$t$, assuming $\omega_0 = 0$.

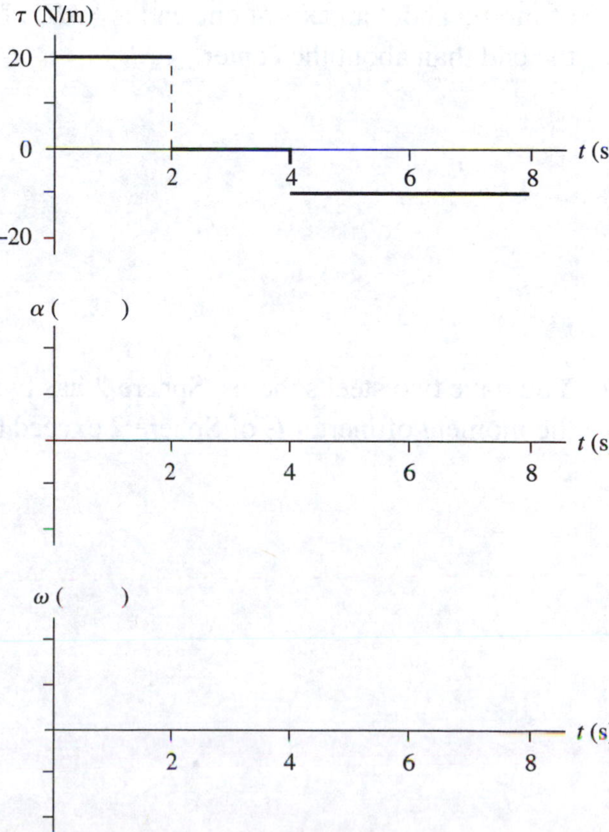

24. The wheel turns on a frictionless axle. A string wrapped around the smaller diameter shaft is tied to a block. The block is released at $t = 0$ s and hits the ground at $t = t_1$.
a. Draw a graph of $\omega$-versus-$t$ for the wheel, starting at $t = 0$ s and continuing to some time $t > t_1$.

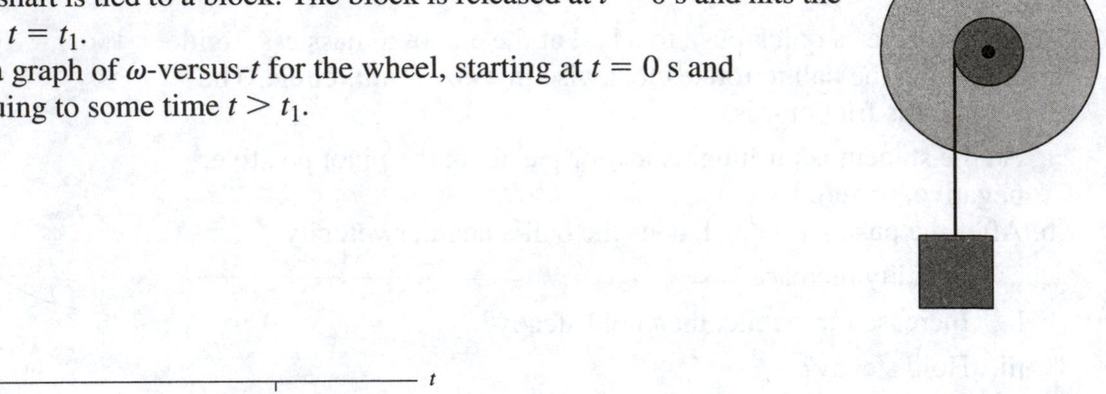

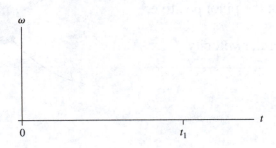

b. Is the magnitude of the block's downward acceleration greater than $g$, less than $g$, or equal to $g$? Explain.

25. The moment of inertia of a uniform rod about an axis through its center is $\frac{1}{12}ML^2$. The moment of inertia about an axis at one end is $\frac{1}{3}ML^2$. Explain *why* the moment of inertia is larger about the end than about the center.

26. You have two steel spheres. Sphere 2 has twice the radius of Sphere 1. By what *factor* does the moment of inertia $I_2$ of Sphere 2 exceed the moment of inertia $I_1$ of Sphere 1?

27. The professor hands you two spheres. They have the same mass, the same radius, and the same exterior surface. The professor claims that one is a solid sphere and that the other is hollow. Can you determine which is which without cutting them open? If so, how? If not, why not?

28. Rank in order, from largest to smallest, the moments of inertia $I_1$, $I_2$, and $I_3$ about the midpoint of each connecting rod.

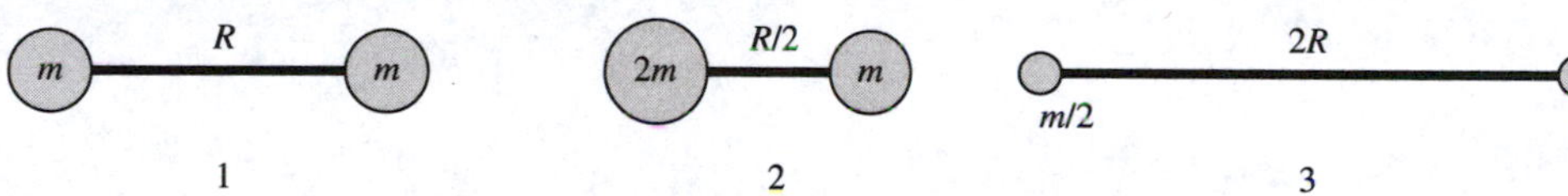


Order:

Explanation:

## 7.6 Using Newton's Second Law for Rotation

29. A square plate can rotate about an axle through its center. Four forces of equal magnitude are applied to different points on the plate. The forces turn as the plate rotates, maintaining the same orientation with respect to the plate. Rank in order, from largest to smallest, the angular accelerations $\alpha_1$ to $\alpha_4$.

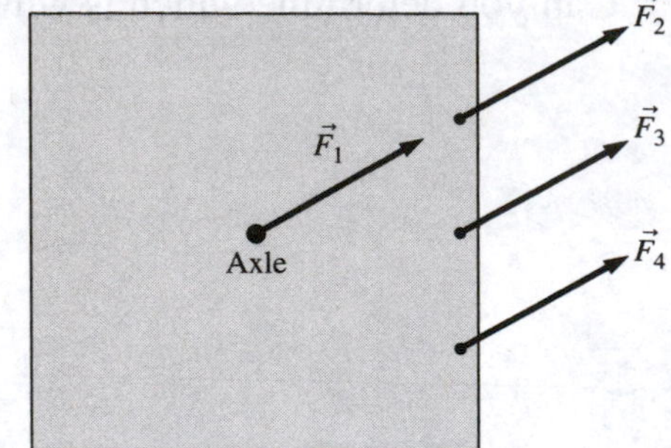


Order:

Explanation:

30. A solid cylinder and a cylindrical shell have the same mass, same radius, and turn on frictionless, horizontal axles. (The cylindrical shell has lightweight spokes connecting the shell to the axle.) A rope is wrapped around each cylinder and tied to a block. The blocks have the same mass and are held the same height above the ground. Both blocks are released simultaneously. The ropes do not slip.

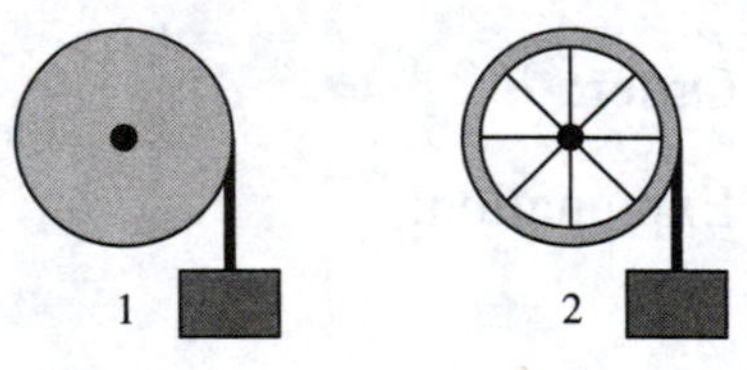


Which block hits the ground first? Or is it a tie? Explain.

31. A metal bar of mass $M$ and length $L$ can rotate in a horizontal plane about a vertical, frictionless axle through its center. A hollow channel down the bar allows compressed air (fed in at the axle) to spray out of two small holes at the ends of the bar, as shown. The bar is found to speed up to angular velocity $\omega$ in a time interval $\Delta t$, starting from rest. What force does each escaping jet of air exert on the bar?

PSS 7.1

Axle

Top view

a. <u>On the figure</u>, draw vectors to show the forces exerted on the bar. Then label the moment arms of each force.

b. The forces in your drawing exert a torque about the axles. Write an expression for each torque, and then add them to get the net torque. Your expression should be in terms of the unknown force $F$ and "known" quantities such as $M$, $L$, $g$, etc.

c. What is the moment of inertia of this bar about the axle? ______________

d. According to Newton's second law, the torque causes the bar to undergo an angular acceleration. Use your results from parts b and c to write an expression for the angular acceleration. Simplify the expression as much as possible.

e. You can now use rotational kinematics to write an expression for the bar's angular velocity after time $\Delta t$ has elapsed. Do so.

f. Finally, solve your equation in part e for the unknown force.

This is now a result you could use with experimental measurements to determine the size of the force exerted by the gas.

## 7.7 Rolling Motion

32. A wheel is rolling along a horizontal surface with the velocity shown. Draw the velocity vectors $\vec{v}_1$ to $\vec{v}_4$ at points 1 to 4 on the rim of the wheel.

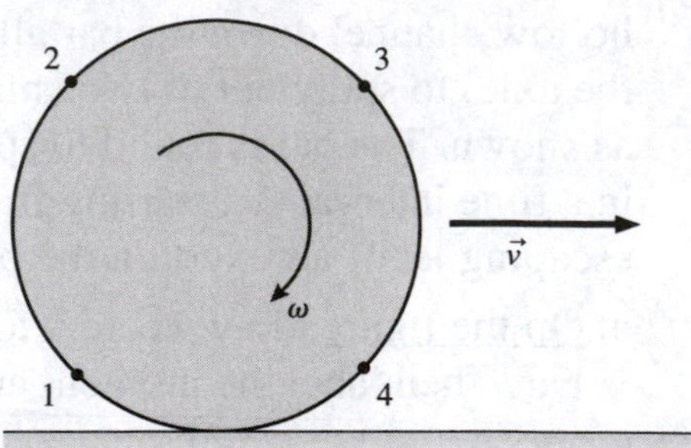


33. A wheel is rolling along a horizontal surface with the velocity shown. Draw the velocity vectors $\vec{v}_1$ to $\vec{v}_3$ at points 1 to 3 on the wheel.

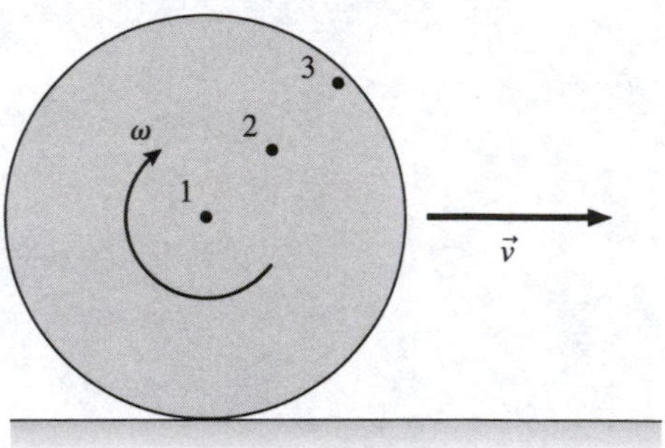


34. If a solid disk and a circular hoop of the same mass and radius are released from rest at the top of a ramp and allowed to roll to the bottom, the disk will get to the bottom first. *Without referring to equations*, explain why this is so.

# 8 Equilibrium and Elasticity

## 8.1 Torque and Static Equilibrium

1. A uniform rod pivots about a frictionless, horizontal axle through its center. It is placed on a stand, held motionless in the position shown, and then gently released. On the right side of the figure, draw the final, equilibrium position of the rod. Explain your reasoning.

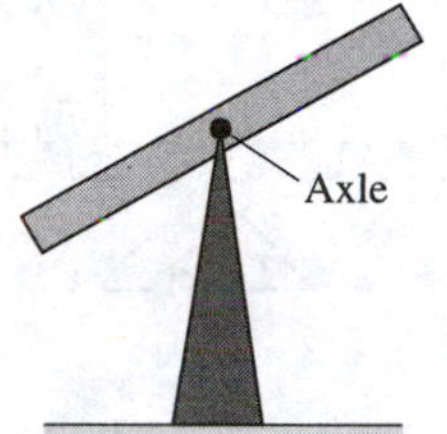


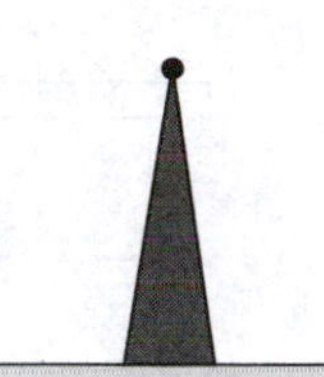

2. A dumbbell consists of masses $m$ and $2m$ connected by a massless, rigid rod. Force $\vec{F}_1$ acts on mass $m$ as shown. Draw and label the vector of a force $\vec{F}_2$ acting on mass $2m$ that will cause the dumbbell to have pure translational motion without any rotation. Make sure the length of your vector shows the magnitude of $\vec{F}_2$ relative to $\vec{F}_1$.

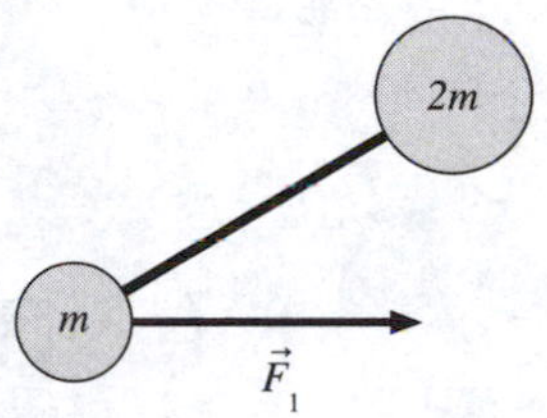


3. Forces $\vec{F}_1$ and $\vec{F}_2$ have the same magnitude. They are applied to the corners of a square plate as shown. Draw and label a *single* force vector $\vec{F}_3$ applied to the appropriate point on the plate that will cause the plate to be in total static equilibrium. Make sure the length of your vector shows the magnitude of $\vec{F}_3$ relative to $\vec{F}_1$ and $\vec{F}_2$.

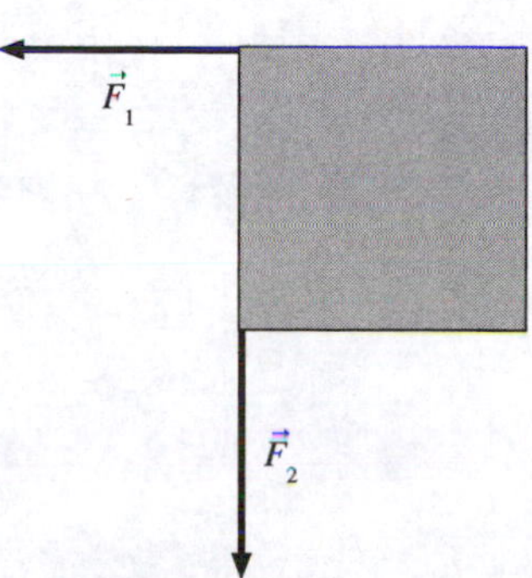

**Exercises 4–7:** Each of the following shows a uniform beam of weight $w$ and several blocks, each of which also has weight $w$. The gravitational force acting downward on the center of gravity of the beam is already shown. For each:

- Draw and label force vectors at each point where a force acts on the beam. You should have both upward and downward vectors. The length of each vector should indicate its magnitude relative to the vector showing the weight of the beam.
- Explain why the beam is or is not in equilibrium.

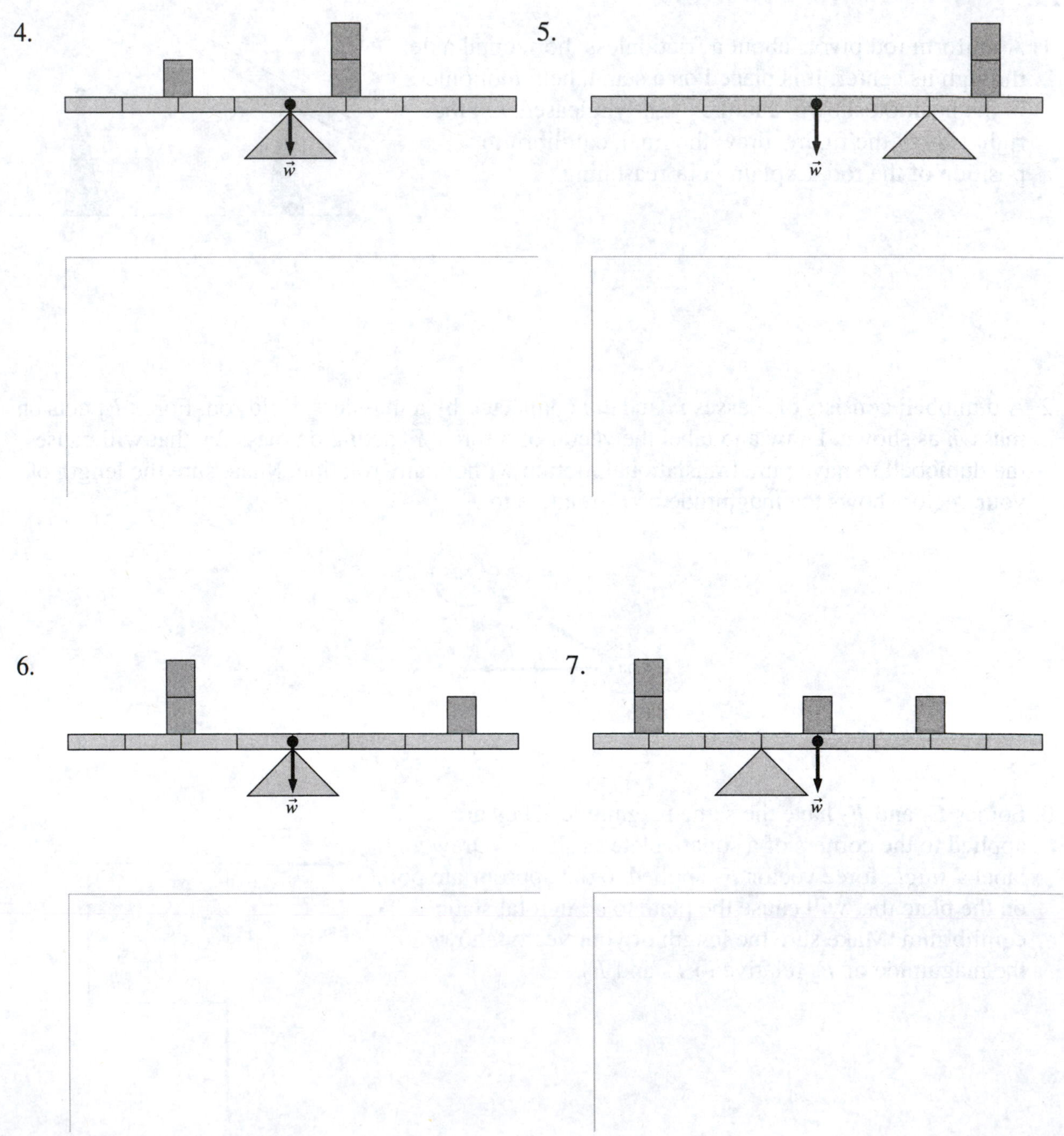

## 8.2 Stability and Balance

8. The center of gravity is marked on each of the objects below.
   - Show and label with a $t$ the object's track width or base of support.
   - Show and label with an $h$ the height of the object's center of gravity.
   - Use a ruler to measure the track width and the height of the center of gravity, and then calculate the critical angle.

a. 

b. 

c. 

| a. | b. | c. |
|---|---|---|
| $t =$ | $t =$ | $t =$ |
| $h =$ | $h =$ | $h =$ |
| $\theta_c =$ | $\theta_c =$ | $\theta_c =$ |

9. Tightrope walkers often carry a long pole that is weighted at the end. The pole serves two purposes: to change the critical angle for balance, and to increase the walker's moment of inertia.
   a. Use the diagrams below to describe how the critical angle is changed by the tightrope walker's use of the pole.

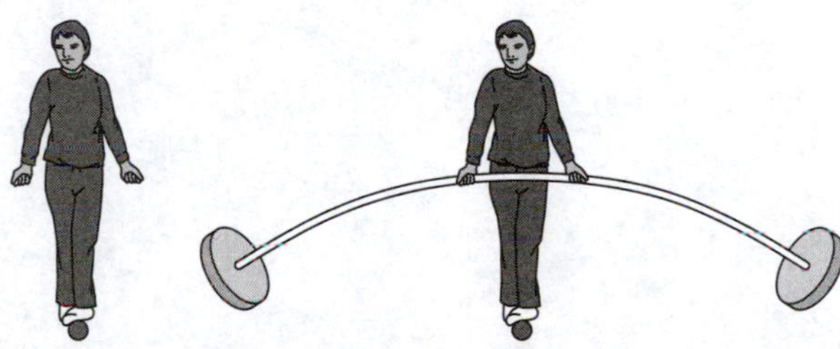

   b. Why would increasing the tightrope walker's moment of inertia also help to make him less likely to fall?

## 8.3 Springs and Hooke's Law

10. The graph below shows the stretching of two different springs, A and B, when different forces were applied.

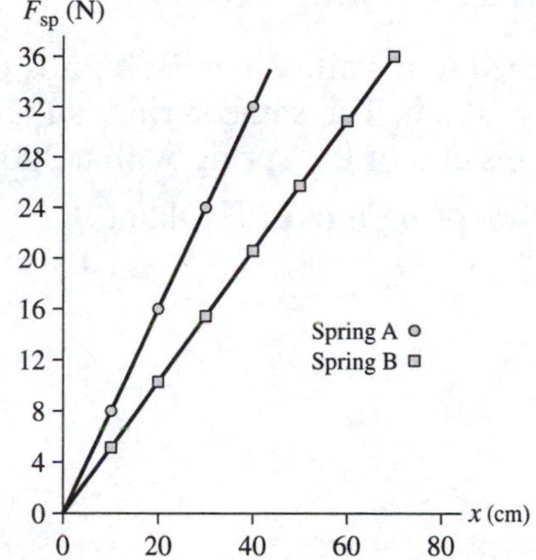


a. Which spring is stiffer? That is, which spring requires a larger pull to get the same amount of stretch? Explain how you can tell by looking at the graph.

b. Determine the spring constant for each spring.

$k_A$ = ____________ $k_B$ = ____________

11. A spring is attached to the floor and pulled straight up by a string. The spring's tension is measured. The graph shows the tension in the spring as a function of the spring's length $L$.

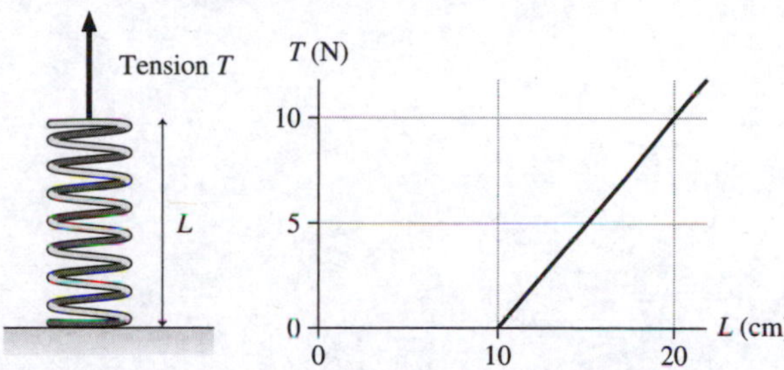


a. Does this spring obey Hooke's Law? Explain why or why not.

b. If it does, what is the spring constant?

12. A spring has an unstretched length of 10 cm. It exerts a restoring force $F$ when stretched to a length of 11 cm.

a. For what length of the spring is its restoring force $3F$? ____________________

b. At what compressed length is the restoring force $2F$? ____________________

13. The left end of a spring is attached to a wall. When Bob pulls on the right end with a 200 N force, he stretches the spring by 20 cm. The same spring is then used for a tug-of-war between Bob and Carlos. Each pulls on his end of the spring with a 200 N force.

a. How far does Bob's end of the spring move? Explain.

b. How far does Carlos's end of the spring move? Explain.

14. A weight hung from a spring stretches the spring by 4.0 cm. If the same weight is hung from a second spring having half the spring constant as the first, by how much will the second spring stretch? Explain.

## 8.4 Stretching and Compressing Materials

15. A force stretches a wire by 1 mm.
    a. A second wire of the same material has the same cross section and twice the length. How far will it be stretched by the same force? Explain.

    b. A third wire of the same material has the same length and twice the diameter as the first. How far will it be stretched by the same force? Explain.

16. A 2000 N force stretches a wire by 1 mm.
    a. A second wire of the same material is twice as long and has twice the diameter. How much force is needed to stretch it by 1 mm? Explain.

    b. A third wire is twice as long as the first and has the same diameter. How far is it stretched by a 4000 N force?

17. A wire is stretched right to the breaking point by a 5000 N force. A longer wire made of the same material has the same diameter. Is the force that will stretch it right to the breaking point larger than, smaller than, or equal to 5000 N? Explain.

## You Write the Problem!

**Exercises 18–19:** You are given the equation that is used to solve a problem. For each of these:

a. Write a *realistic* physics problem for which this is the correct equation. Look at worked examples and end-of-chapter problems in the textbook to see what realistic physics problems are like. Be sure that the problem you write, and the answer you ask for, is consistent with the information given in the equation.
b. Draw a free-body diagram or force diagram for your problem.
c. Finish the solution of the problem.

18. $n + T - 80\text{ N} = 0$
$(0.60\text{ m})\,T - (0.50\text{ m})(80\text{ N}) = 0$

19. $k(0.050\text{ m}) - (0.15)(0.12\text{ kg})(9.8\text{ m/s}^2) = (0.12\text{ kg})(0.65\text{ m/s}^2)$

# 9 Momentum

## 9.1 Impulse

## 9.2 Momentum and the Impulse-Momentum Theorem

1. What impulse is delivered by each of these forces?

a. $F$ (N) 5 0 −5 0.2 s $t$ (s)

b.

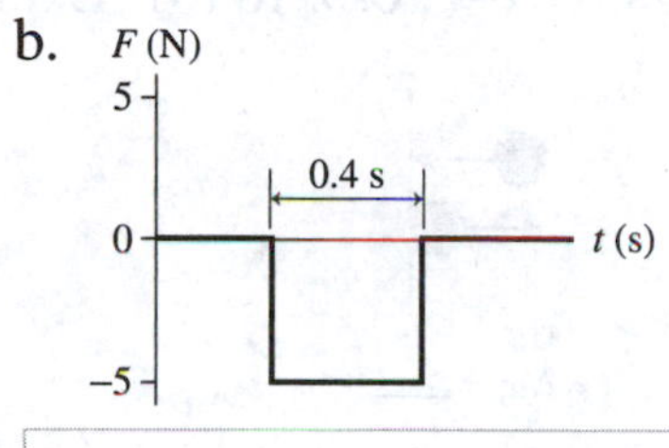


c.

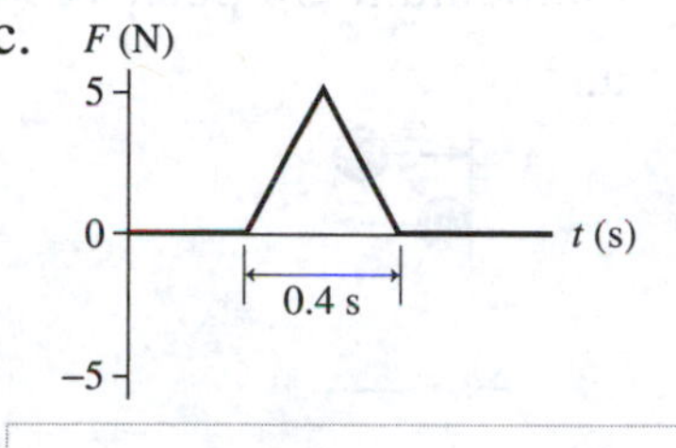


2. Rank in order, from largest to smallest, the momenta $(p_x)_1$ to $(p_x)_5$.

20 g 1 1 m/s

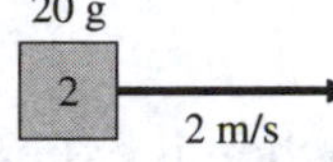


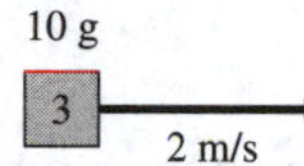


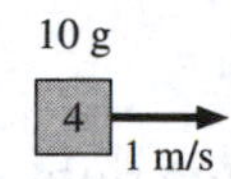


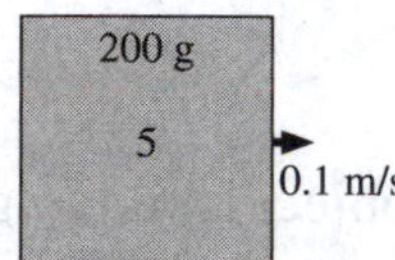


Order:

Explanation:

3. The position-versus-time graph is shown for a 500 g object. Draw the corresponding momentum-versus-time graph. Include an appropriate vertical scale.

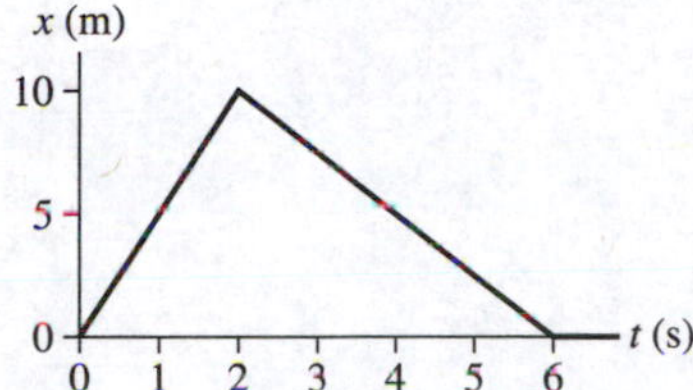


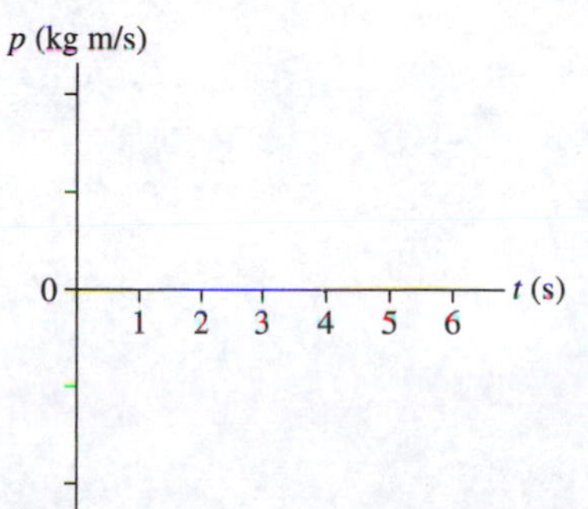

4. The momentum-versus-time graph is shown for a 500 g object. Draw the corresponding acceleration-versus-time graph. Include an appropriate vertical scale.

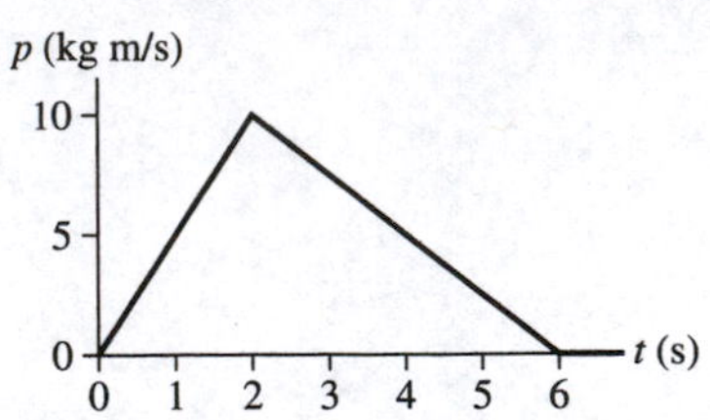


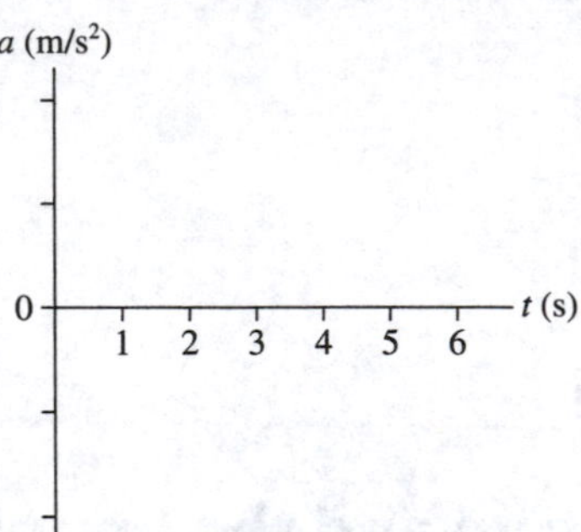


5. In each of the following, where a rubber ball bounces with no loss of speed, is the change in momentum $\Delta p$ positive (+), negative (–) , or zero (0)? Explain.

a.

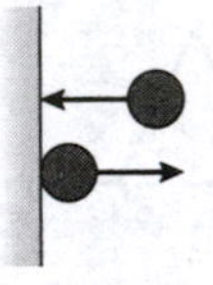

$\Delta p_x =$ ____

b.

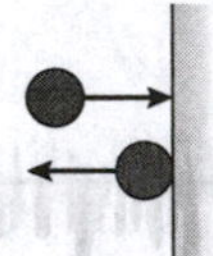

$\Delta p_x =$ ____

c.

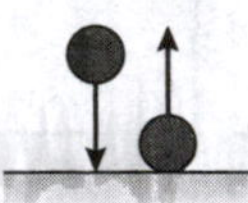

$\Delta p_y =$ ____

6. A 2 kg object is moving to the right with a speed of 1 m/s when it experiences an impulse due to the force shown in the graph. What is the object's speed and direction after the impulse?

a.

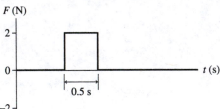

b.

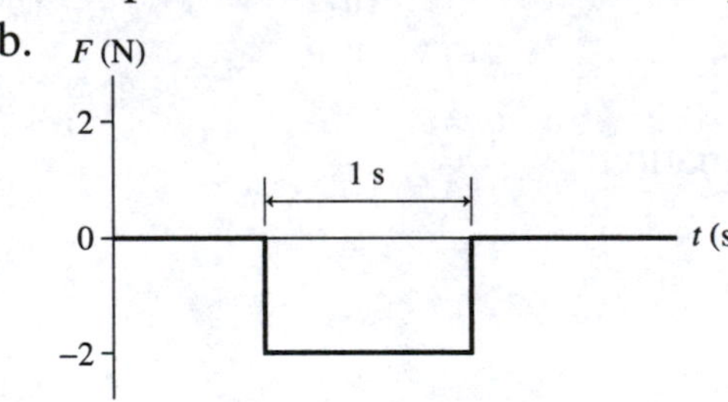

7. A carnival game requires you to knock over a wood post by throwing a ball at it. You're offered a very bouncy rubber ball and a very sticky clay ball of equal mass. Assume that you can throw them with equal speed and equal accuracy. You only get one throw.

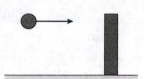

   a. Which ball will you choose? Why?

   b. Let's think about the situation more carefully. Both balls have the same initial momentum $(p_x)_i$ just before hitting the post. The clay ball sticks, the rubber ball bounces off with essentially no loss of speed. What is the final momentum of each ball?

      Clay ball: $(p_x)_f =$ ______ Rubber ball: $(p_x)_f =$ ______

      Hint: Momentum has a sign. Did you take the sign into account?

   c. What is the *change* in the momentum of each ball?

      Clay ball: $\Delta p_x =$ ______ Rubber ball: $\Delta p_x =$ ______

   d. Which ball experiences a larger impulse during the collision? Explain.

   e. From Newton's third law, the impulse that the ball exerts on the post is equal in magnitude, although opposite in direction, to the impulse that the post exerts on the ball. Which ball exerts the larger impulse on the post?

   f. Don't change your answer to part a, but are you still happy with that answer? If not, how would you change your answer? Why?

8. A small, light ball S and a large, heavy ball L move toward each other, collide, and bounce apart.

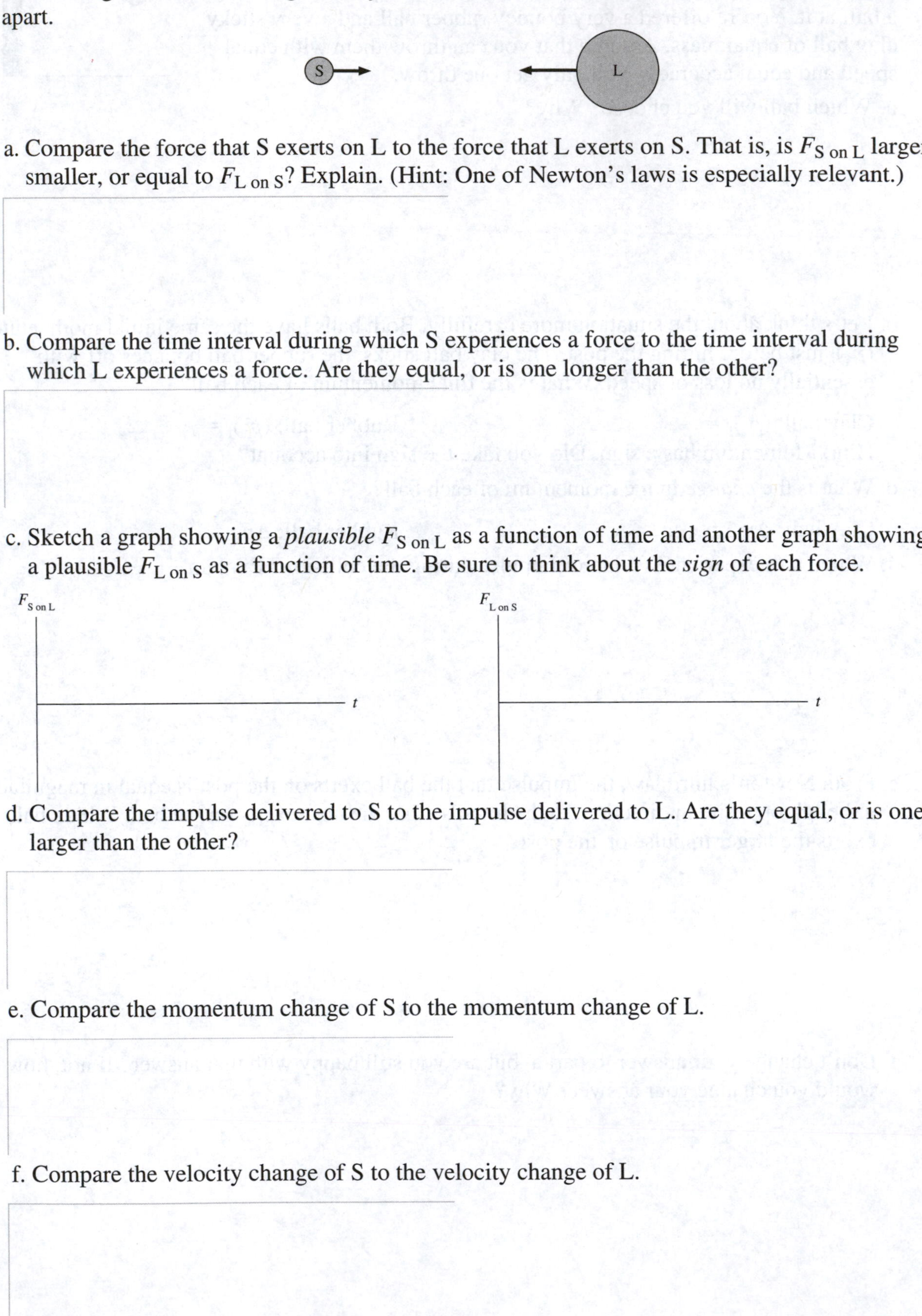


a. Compare the force that S exerts on L to the force that L exerts on S. That is, is $F_{\text{S on L}}$ larger, smaller, or equal to $F_{\text{L on S}}$? Explain. (Hint: One of Newton's laws is especially relevant.)

b. Compare the time interval during which S experiences a force to the time interval during which L experiences a force. Are they equal, or is one longer than the other?

c. Sketch a graph showing a *plausible* $F_{\text{S on L}}$ as a function of time and another graph showing a plausible $F_{\text{L on S}}$ as a function of time. Be sure to think about the *sign* of each force.

d. Compare the impulse delivered to S to the impulse delivered to L. Are they equal, or is one larger than the other?

e. Compare the momentum change of S to the momentum change of L.

f. Compare the velocity change of S to the velocity change of L.

## 9.3 Solving Impulse and Momentum Problems

**Exercises 9–11:** Prepare a before-and-after visual overview for these problems, but *do not* solve them.

- Draw pictures of "before" and "after."
- Establish a coordinate system.
- Define symbols relevant to the problem.
- List known information, *and* identify the desired unknown.

9. A 50 kg archer, standing on frictionless ice, shoots a 100 g arrow at a speed of 100 m/s. What is the recoil speed of the archer?

10. The parking brake on a 2000 kg Cadillac has failed, and it is rolling slowly, at 1 mph, toward a group of small innocent children. As you see the situation, you realize there is just time for you to drive your 1000 kg Volkswagen head-on into the Cadillac and thus save the children. With what speed should you impact the Cadillac to bring it to a halt?

11. Dan is gliding on his skateboard at 4 m/s. He suddenly jumps backward off the skateboard, kicking the skateboard forward at 8 m/s. How fast is Dan going as his feet hit the ground? Dan's mass is 50 kg and the skateboard's mass is 5 kg.

12. Blocks A and B, both initially at rest, are pushed to the right continuously by identical constant forces. Block B is more massive than Block A. Which block crosses the finish line with more momentum? Or do they finish with equal momenta? Explain.

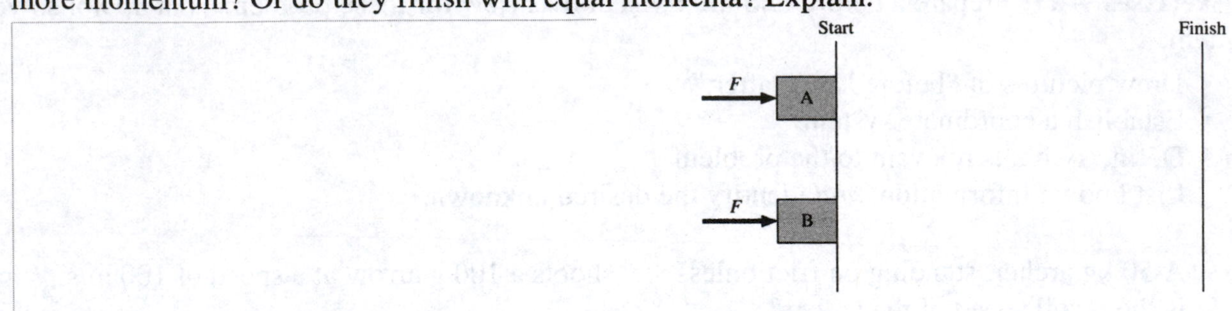


13. Blocks A and B, of equal mass, are pushed to the right continuously by identical constant forces. Block B starts from rest, but Block A is already moving to the right as it crosses the starting line. Which block undergoes a larger *change* in momentum before crossing the finish line? Or are they the same? Explain.

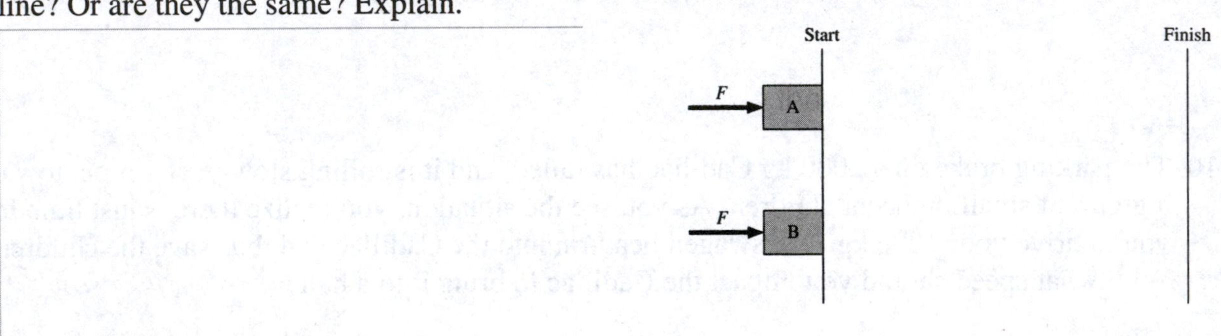


14. Blocks A and B are pushed to the right continuously by identical constant forces for exactly 1.0 s, starting from rest. Block B is more massive than Block A. After 1.0 s, which block has more momentum? Or do they have equal momenta? Explain.

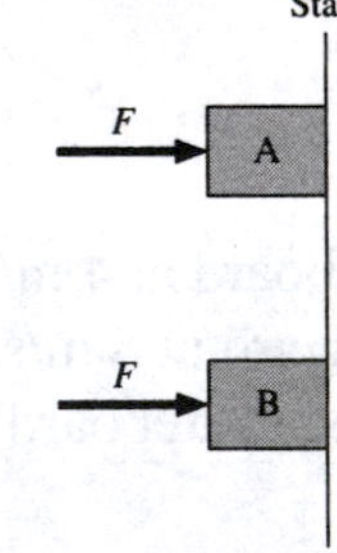

## 9.4 Conservation of Momentum

15. As you release a ball, it falls—gaining speed and momentum. Is momentum conserved?
    a. Answer this question from the perspective of choosing the ball alone as the system.

    b. Answer this question from the perspective of choosing ball + earth as the system.

16. Two particles collide, one of which was initially moving and the other initially at rest.
    a. Is it possible for *both* particles to be at rest after the collision? Give an example in which this happens, or explain why it can't happen.

    b. Is it possible for *one* particle to be at rest after the collision? Give an example in which this happens, or explain why it can't happen.

17. A tennis ball traveling to the left at speed $v_{Bi}$ is hit by a tennis racket moving to the right at speed $v_{Ri}$. Although the racket is swung in a circular arc, its forward motion during the collision with the ball is so small that we can consider it to be moving in a straight line. Further, we can invoke the *impulse approximation* to neglect the steady force of the arm on the racket during the brief duration of its collision with the ball. Afterward, the ball is returned to the right at speed $v_{Bf}$. What is the racket's speed after it hits the ball? The masses of the ball and racket are $m_B$ and $m_R$, respectively.

PSS 9.1

a. Begin by drawing a before-and-after visual overview, as described in Tactics Box 9.1. You can assume that the racket continues in the forward direction but at a reduced speed.

b. Define the system. That is, what object or objects should be inside the system so that it is an *isolated system* whose momentum is conserved?

c. Write an expression for $(P_x)_i$, the total momentum of the system before the collision. Your expression should be written using the quantities given in the problem statement. Notice, however, that you're given *speeds*, but momentum is defined in terms of *velocities*. Based on your coordinate system and the directions of motion, you may need to give a negative momentum to one or more objects.

d. Now write an expression for $(P_x)_f$, the total momentum of the system after the collision.

e. If you chose the system correctly, its momentum is conserved. So equate your expressions for the initial and final total momentum, and then solve for what you want to find.

## 9.5 Inelastic Collisions

18. Determine whether each of the following graphs represents an inelastic collision between Object A (solid line) and Object B (dashed line). The objects in an inelastic collision must stick together, *and* the collision must conserve momentum. Part of your explanation should consider the relative masses of A and B.

    Note that part a is a position-versus-time graph, but parts b–d are velocity-versus-time graphs.

a.

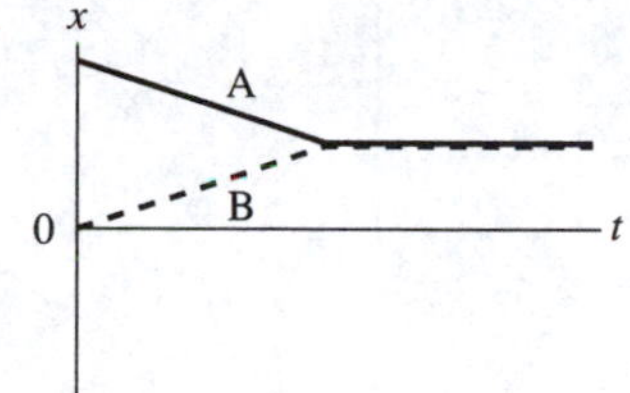


b.

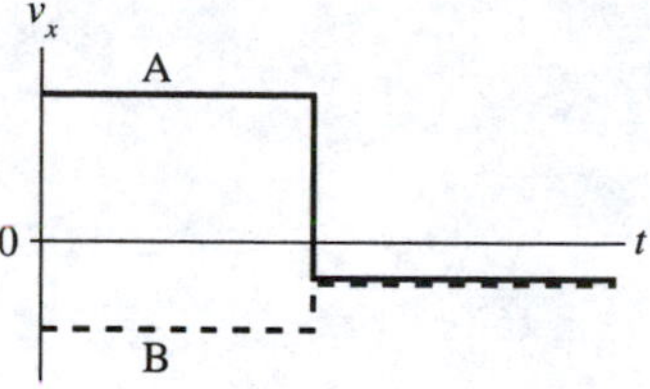


c.

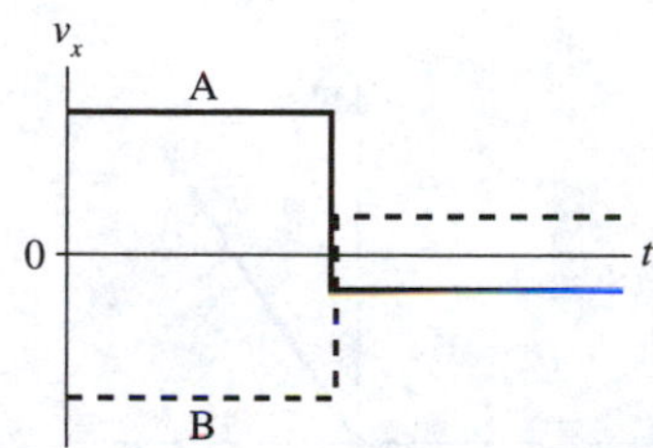


d.

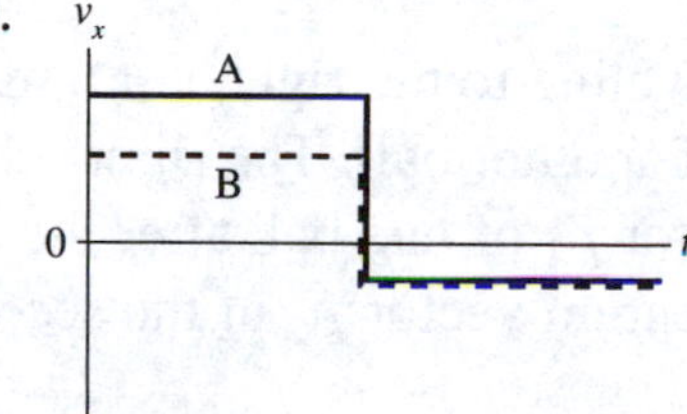

## 9.6 Momentum and Collisions in Two Dimensions

19. An object initially at rest explodes into three fragments. The momentum vectors of two of the fragments are shown. Draw the momentum vector $\vec{p}_3$ of the third fragment.

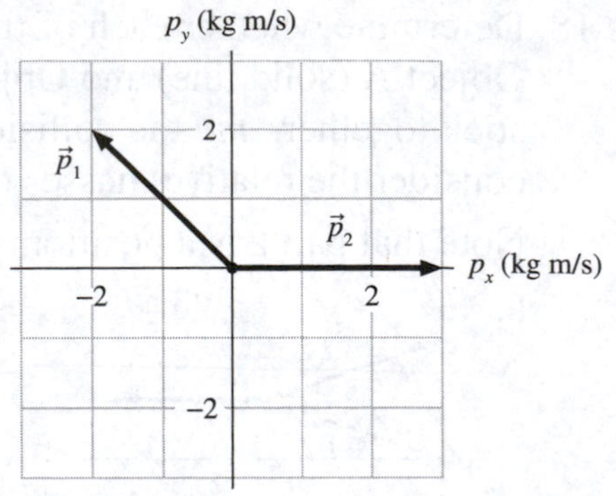


20. An object initially at rest explodes into three fragments. The momentum vectors of two of the fragments are shown. Draw the momentum vector $\vec{p}_3$ of the third fragment.

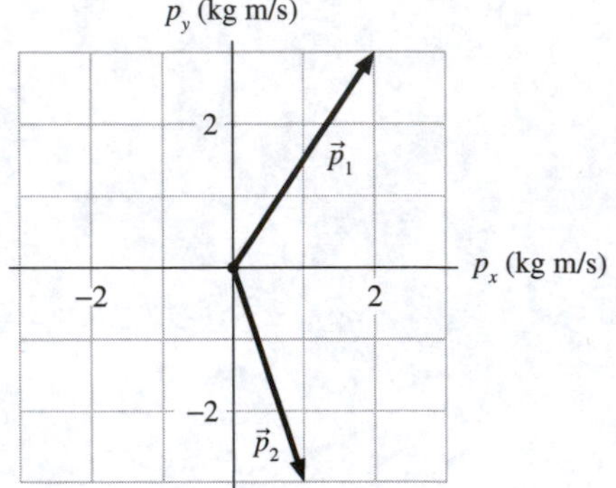


21. A 500 g ball traveling to the right at 4.0 m/s collides with and bounces off another ball. The figure shows the momentum vector $\vec{p}_1$ of one ball after the collision. Draw the momentum vector $\vec{p}_2$ of the second ball.

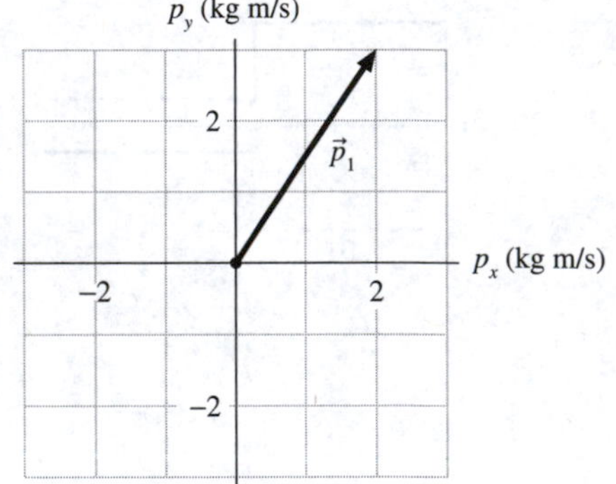


22. A 500 g ball traveling to the right at 4.0 m/s collides with and bounces off another ball. The figure shows the momentum vector $\vec{p}_1$ of one ball after the collision. Draw the momentum vector $\vec{p}_2$ of the other ball.

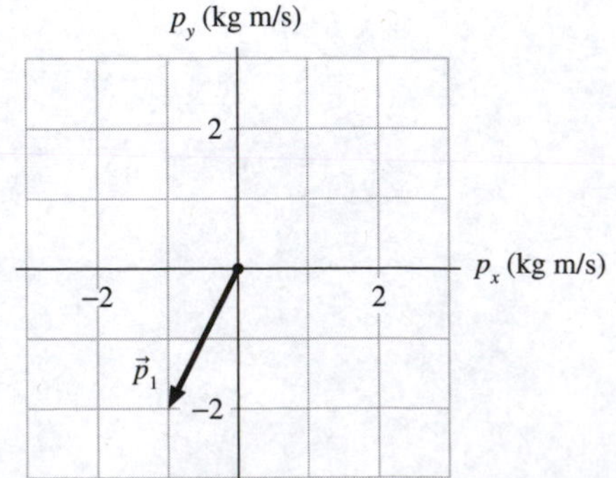

## 9.7 Angular Momentum

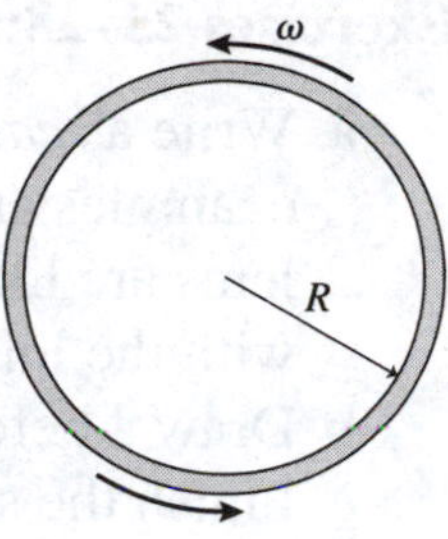


23. An isolated hoop of mass $M$ and radius $R$ is rotating with an angular speed of 60 rpm about its axis.
    a. What would be its angular speed if its mass suddenly doubled without changing its radius? Explain.

    b. What would be its angular speed if its radius suddenly doubled without changing its mass? Explain.

24. A solid circular disk and a circular hoop (like a bicycle wheel) both have mass $M$ and radius $R$. If both are rotating with the same angular velocity $\omega$, which has the larger angular momentum? Or are they equal? Explain.

## You Write the Problem!

**Exercises 25–28:** You are given the equation that is used to solve a problem. For each of these:

a. Write a *realistic* physics problem for which this is the correct equation. Look at worked examples and end-of-chapter problems in the textbook to see what realistic physics problems are like. Be sure that the problem you write, and the answer you ask for, is consistent with the information given in the equation.
b. Draw a before-and-after visual overview for your problem.
c. Finish the solution of the problem.

25. $(0.10\text{ kg})(40\text{ m/s}) - (0.10\text{ kg})(-30\text{ m/s}) = \frac{1}{2}(1400\text{ N})\,\Delta t$

26. $(600\text{ g})(4.0\text{ m/s}) = (400\text{ g})(3.0\text{ m/s}) + (200\text{ g})(v_{2x})_i$

27. $(3000\text{ kg})(v_x)_f = (2000\text{ kg})(5.0\text{ m/s}) + (1000\text{ kg})(-4.0\text{ m/s})$

28. $(50\text{ g})(v_{1x})_f + (100\text{ g})(7.5\text{ m/s}) = (150\text{ g})(1.0\text{ m/s})$

# 10 Energy and Work

## 10.1 The Basic Energy Model

1. What are the two primary processes by which energy can be transferred from the environment to a system?

2. Identify the energy *transformations* in each of the following processes (e.g., $K \rightarrow U_g \rightarrow E_{th}$)
   a. A ball is dropped from atop a tall building.

   b. A helicopter rises from the ground at constant speed.

   c. An arrow is shot from a bow and stops in the center of its target.

   d. A pole vaulter runs, plants his pole, and vaults up over the bar.

3. Identify the energy *transfers* that occur in each of the following processes (e.g., $W \rightarrow K$).
   a. You pick up a book from the floor and place it on a table.

   b. You roll a bowling ball.

   c. You sand a board with sandpaper, making the board and the sandpaper warm.

## 10.2 Work

4. For each situation described below:
   - Draw a before-and-after diagram, similar to Figures 10.8 and 10.12 in the textbook.
   - Identify *all* forces acting on the object.
   - Determine if the work done by each of these forces is positive (+), negative (−), or zero (0). Make a little table beside the figure showing *every* force and the sign of its work.

   a. An elevator moves upward.

   b. An elevator moves downward.

   c. You slide down a steep hill.

5. An object experiences a force while undergoing the displacement shown. Is the work done positive (+), negative (−), or zero (0)?

a.

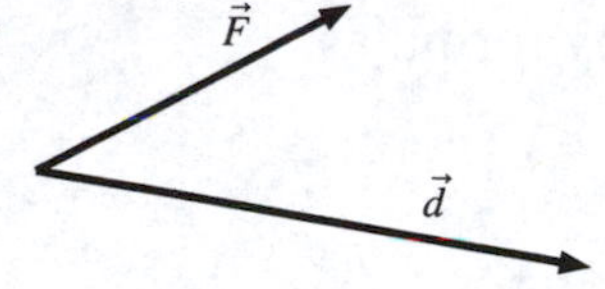


Sign of $W$ = ________

b.

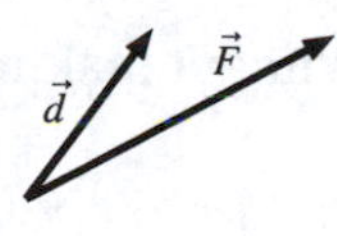


Sign of $W$ = ________

c.

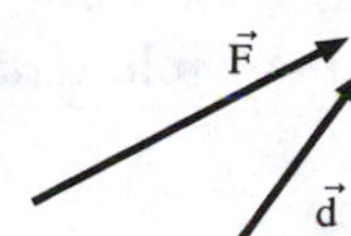


Sign of $W$ = ________

d.

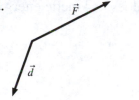

Sign of $W$ = ________

e.

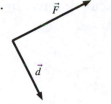

Sign of $W$ = ________

f.

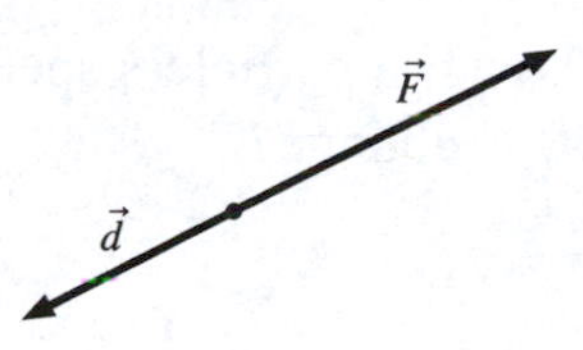


Sign of $W$ = ________

6. Each of the diagrams below shows a displacement vector for an object. Draw and label a force vector that will do work on the object with the sign indicated.

a.

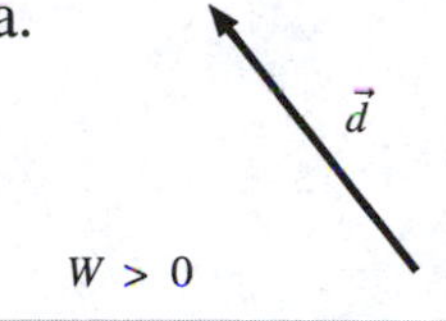


$W > 0$

b.

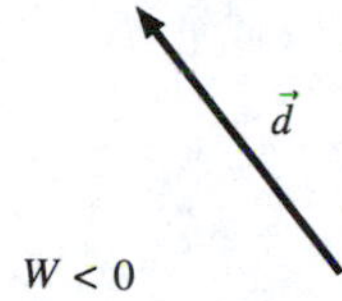


$W < 0$

c.

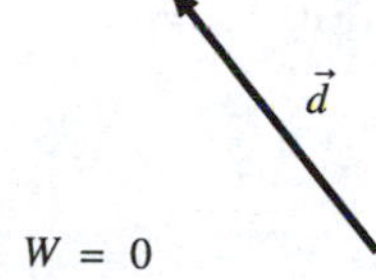


$W = 0$

7. A 0.2 kg plastic cart and a 20 kg lead cart both roll without friction on a horizontal surface. Equal forces are used to push both carts forward a distance of 1 m, starting from rest. After traveling 1 m, is the kinetic energy of the plastic cart greater than, less than, or equal to the kinetic energy of the lead cart? Explain.

## 10.3 Kinetic Energy

8. Can kinetic energy ever be negative? ____________

Give a plausible *reason* for your answer without making use of any formulas.

9. a. If a particle's speed increases by a factor of three, by what factor does its kinetic energy change?

b. Particle A has half the mass and eight times the kinetic energy of particle B. What is the speed ratio $v_A/v_B$?

10. On the axes below, draw graphs of the kinetic energy of

a. A 1000 kg car that uniformly accelerates from 0 to 20 m/s in 20 s.
b. A 1000 kg car moving at 20 m/s that brakes to a halt with uniform deceleration in 4 s.
c. A 1000 kg car that drives once around a 40-m-diameter circle at a speed of 20 m/s.

Calculate $K$ at several times, plot the points, and draw a smooth curve between them.

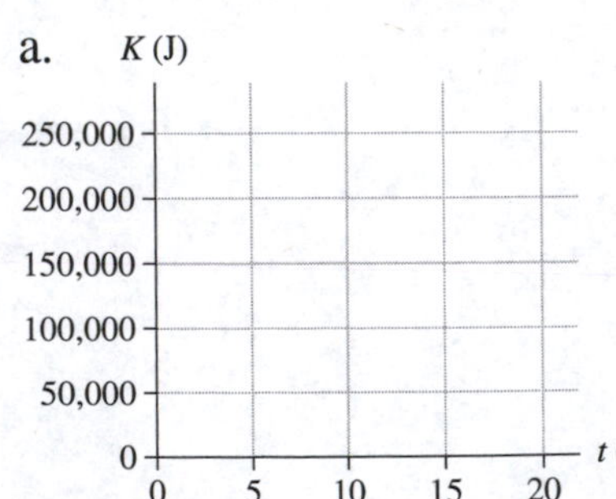


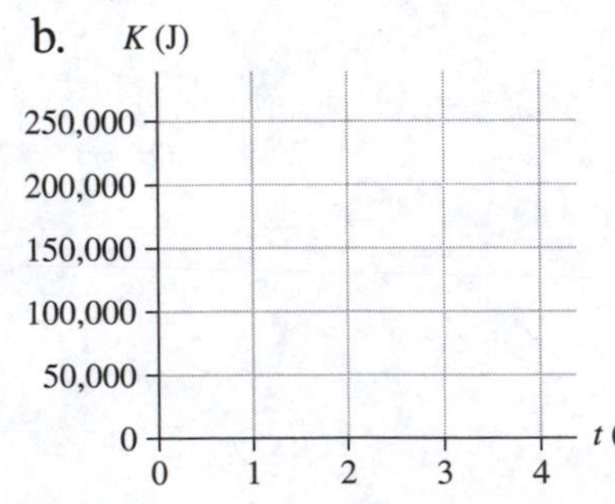


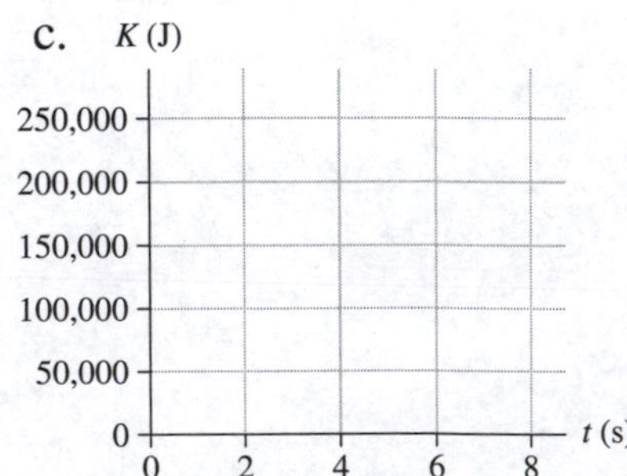

## 10.4 Potential Energy

11. Below we see a 1 kg object that is initially 1 m above the ground and rises to a height of 2 m. Anjay and Brittany each measure its position but use a different coordinate system to do so. Fill in the table to show the initial and final gravitational potential energies and $\Delta U_g$ as measured by Anjay and Brittany.

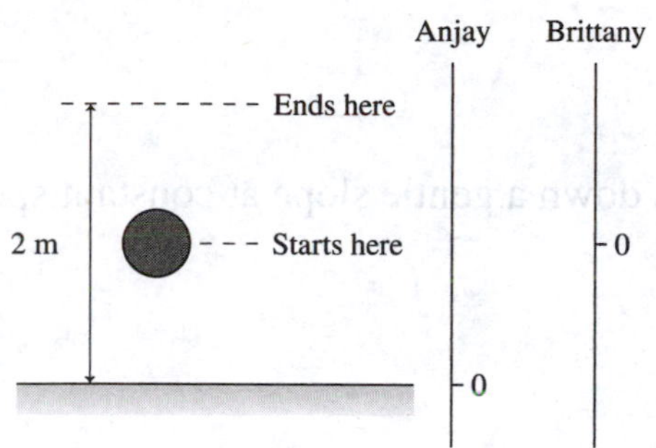


| | $(U_g)_i$ | $(U_g)_f$ | $\Delta U_g$ |
|---|---|---|---|
| Anjay | | | |
| Brittany | | | |

12. Rank in order, from most to least, the amount of elastic potential energy $(U_s)_1$ to $(U_s)_4$ stored in each of these springs.

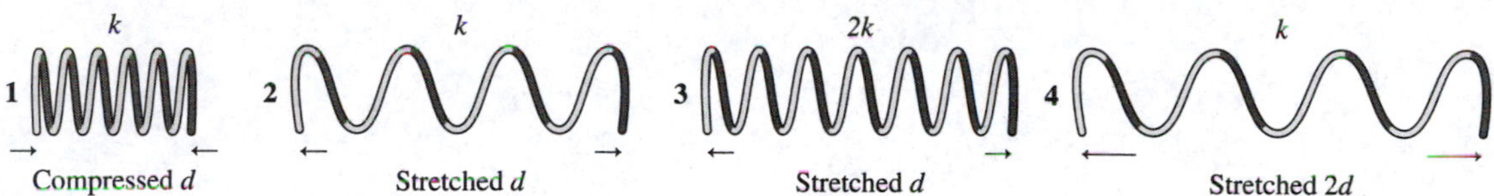


Order:

Explanation:

13. A heavy object is released from rest at position 1 above a spring. It falls and contacts the spring at position 2. The spring achieves maximum compression at position 3. Fill in the table below to indicate whether each of the quantities are +, −, or 0 during the intervals 1→2, 2→3, and 1→3.

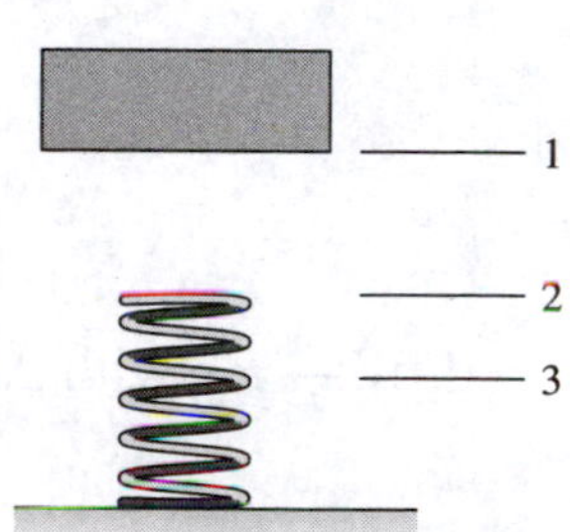


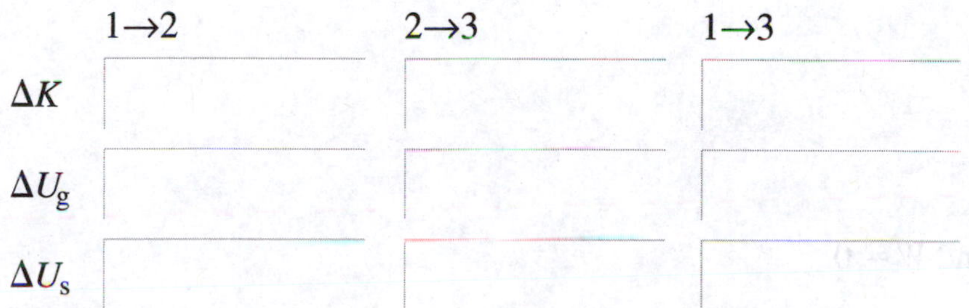

| | 1→2 | 2→3 | 1→3 |
|---|---|---|---|
| $\Delta K$ | | | |
| $\Delta U_g$ | | | |
| $\Delta U_s$ | | | |

## 10.5 Thermal Energy

14. A car traveling at 60 mph slams on its brakes and skids to a halt. What happened to the kinetic energy the car had just before stopping?

15. What energy transformations occur as a skier glides down a gentle slope at constant speed?

16. Give a *specific* example of a situation in which:

a. $W \rightarrow K$ with $\Delta U = 0$ and $\Delta E_{th} = 0$.

b. $W \rightarrow U$ with $\Delta K = 0$ and $\Delta E_{th} = 0$.

c. $K \rightarrow U$ with $W = 0$ and $\Delta E_{th} = 0$.

d. $W \rightarrow E_{th}$ with $\Delta K = 0$ and $\Delta U = 0$.

e. $U \rightarrow E_{th}$ with $\Delta K = 0$ and $W = 0$.

## 10.6 Using the Law of Conservation of Energy

17. What is meant by an *isolated system*?

18. Identify an appropriate system for applying conservation of energy to each of the following:

a. A spring is used to launch a ball into the air.

System:

b. A spring is used to push a car on an air track.

System:

c. A spring is used to slide a block across a table where it stops.

System:

d. A car moving on an air track collides with a spring and rebounds at essentially the same speed with which it hit the spring.

System:

19. a. A process occurs in which a system's potential energy decreases while the environment does work on the system. Does the system's kinetic energy increase, decrease, or stay the same? Or is there not enough information to tell? Explain.

b. A process occurs in which a system's potential energy increases while the environment does work on the system. Does the system's kinetic energy increase, decrease, or stay the same? Or is there not enough information to tell? Explain.

20. Three balls of equal mass are fired simultaneously with *equal* speeds from the same height above the ground. Ball 1 is fired straight up, Ball 2 is fired straight down, and Ball 3 is fired horizontally. Rank in order, from largest to smallest, their speeds $v_1$, $v_2$, and $v_3$ as they hit the ground.

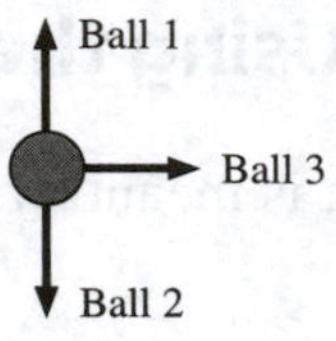


Order:

Explanation:

21. Below are shown three frictionless tracks. A block is released from rest at the position shown on the left. To which point does the block make it on the right before reversing direction and sliding back? Point B is the same height as the starting position.

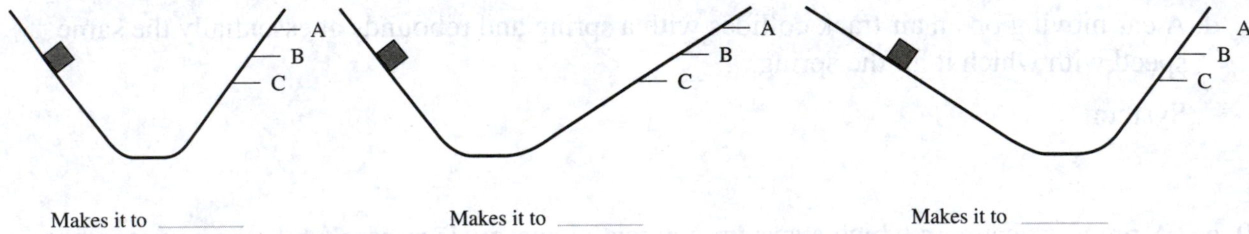


Makes it to ______ Makes it to ______ Makes it to ______

22. A spring gun shoots out a plastic ball at speed $v_0$. The spring is then compressed twice the distance it was on the first shot.

a. By what factor is the spring's potential energy increased?

b. By what factor is the ball's launch speed increased? Explain.

23. A small cube of mass $m$ slides back and forth in a frictionless, hemispherical bowl of radius $R$. Suppose the cube is released at angle $\theta$. What is the cube's speed at the bottom of the bowl?

PSS 10.1

a. Begin by drawing a before-and-after visual overview. Let the cube's initial position and speed be $y_i$ and $v_i$. Use a similar notation for the final position and speed.

b. At the initial position, are either $K_i$ or $(U_g)_i$ zero? If so, which? ______

c. At the final position, are either $K_f$ or $(U_g)_f$ zero? If so, which? ______

d. Does thermal energy need to be considered in this situation? Why or why not?

e. Write the conservation of energy equation in terms of position and speed variables, omitting any terms that are zero.

f. You're given not the initial position but the initial angle. Do the geometry and trigonometry to find $y_i$ in terms of $R$ and $\theta$.

g. Use your result of part f in the energy conservation equation, and then finish solving the problem.

## You Write the Problem!

**Exercises 24–26:** You are given the equation that is used to solve a problem. For each of these:

a. Write a *realistic* physics problem for which this is the correct equation. Look at worked examples and end-of-chapter problems in the textbook to see what realistic physics problems are like. Be sure that the problem you write, and the answer you ask for, is consistent with the information given in the equation.

b. Draw a before-and-after visual overview for your problem.

c. Finish the solution of the problem.

24. $\frac{1}{2}(1500\text{ kg})(5.0\text{ m/s})^2 + (1500\text{ kg})(9.8\text{ m/s}^2)(10\text{ m})$
$= \frac{1}{2}(1500\text{ kg})(v_\text{i})^2 + (1500\text{ kg})(9.8\text{ m/s}^2)(0\text{ m})$

25. $\frac{1}{2}(0.20\text{ kg})(2.0\text{ m/s})^2 + \frac{1}{2}k(0\text{ m})^2 = \frac{1}{2}(0.20\text{ kg})(0\text{ m/s})^2 + \frac{1}{2}k(-0.15\text{ m})^2$

26. $(0.10\text{ kg} + 0.20\text{ kg})(v_{1x}) = (0.10\text{ kg})(3.0\text{ m/s})$
$\frac{1}{2}(0.30\text{ kg})(0\text{ m/s})^2 + \frac{1}{2}(3.0\text{ N/m})(x_2)^2 = \frac{1}{2}(0.30\text{ kg})(v_{1x})^2 + \frac{1}{2}(3.0\text{ N/m})(0\text{ m})^2$

## 10.7 Energy in Collisions

27. Ball 1 with an initial speed of 14 m/s has a perfectly elastic collision with Ball 2 that is initially at rest. Afterward, the speed of Ball 2 is 21 m/s.
   a. What will be the speed of Ball 2 if the initial speed of Ball 1 is doubled?

   b. What will be the speed of Ball 2 if the mass of Ball 1 is doubled?

28. Consider a perfectly elastic collision in which a moving ball 1 strikes an initially stationary Ball 2.
   a. Under what circumstances, if any, will Ball 1 come to a stop?

   b. Under what circumstances, if any, will Ball 1 recoil backwards?

   c. Under what circumstances, if any, will Ball 1 continue moving forward?

29. You can dive into a swimming pool of water from a high diving board without being hurt, but to dive into an empty pool from a much lower distance might be fatal. Why the difference?

## 10.8 Power

30. a. If you push an object 10 m with a 10 N force in the direction of motion, how much work do you do on it?

b. How much power must you provide to push the object in 1 s? In 10 s? In 0.1 s?

31. a. To push an object *twice as fast* with the same force, must your power output increase? If so, by what factor? Explain.

b. To push an object *twice as far* with the same force and at the same speed, must your power output increase? If so, by what factor? Explain.

# 11 Using Energy

## 11.1 Transforming Energy

1. a. Given that efficiency is defined as $e =$ (what you get)/(what you had to pay), how might one define the concept of "inefficiency"?

   b. Write a mathematical relationship to relate your definition of inefficiency, which you can denote as $i$, to the definition of efficiency $e$.

2. A device uses 500 J of chemical energy to generate 100 J of electric energy.
   a. What is the efficiency of this device?

   b. A second device is twice as efficient. How much electric energy can it generate from the same 500 J of chemical energy?

3. Is it possible for the efficiency of some device or machine to be greater than 1? Either give an example where the efficiency is greater than 1 or explain why it's not possible.

4. A small battery-operated car zips across the floor. In thinking about the efficiency of the car:
   a. What do you get? In particular, what useful form of energy increases?

   b. What do you pay? In particular, what form of energy decreases?

## 11.2 Energy in the Body

5. Food contains energy that your body can use. Is the energy in food kinetic, potential, thermal, chemical, or nuclear? Explain.

6. Can you run further by burning an ounce of fat or an ounce of carbohydrates? Explain.

7. The energy used by a human while running is approximately proportional to the speed. If we expressed the energy used by a runner in terms of, say, miles per candy bar, who gets better mileage per candy bar, a runner who completes a marathon in $2\frac{1}{2}$ hours or one who completes the same race in 4 hours? Or is it the same for both runners? Explain.

8. Suppose your body burns 10 Cal climbing stairs. Which of the following will allow you to burn 20 Cal?
   a. Climb twice as high at the same speed.
   b. Climb the original stairs twice as fast.
   c. Either a or b.
   d. Neither a nor b.

   Explain.

## 11.3 Temperature, Thermal Energy, and Heat

9. Rank in order, from highest to lowest, the temperatures $T_1 = 0$ K, $T_2 = 0$°C, and $T_3 = 0$°F.

10. "Room temperature" is often considered to be 68°F. What is room temperature in °C and in K?

11. a. What is the average kinetic energy of atoms at absolute zero?
    b. Can an atom have negative kinetic energy?
    c. Is it possible to have a temperature less than absolute zero? Explain.

12. Do each of the following describe a property of a system, an interaction of a system with its environment, or both? Explain.
    a. Temperature:

    b. Heat:

    c. Thermal energy:

## 11.4 The First Law of Thermodynamics

13. For each of the following processes:
    a. Is the value of the work $W$, the heat $Q$, and the change of thermal energy $\Delta E_{th}$ positive (+), negative (−), or zero (0)?
    b. Does the temperature increase (+), decrease (−), or not change (0)?

| | $W$ | $Q$ | $\Delta E_{th}$ | $\Delta T$ |
|---|---|---|---|---|
| • You hit a nail with a hammer. | | | | |
| • You hold a nail over a Bunsen burner. | | | | |
| • High-pressure steam spins a turbine. | | | | |
| • Steam contacts a cold surface and condenses. | | | | |
| • A moving crate slides to a halt on a rough surface. | | | | |

## 11.5 Heat Engines

14. Rank in order, from largest to smallest, the efficiencies $e_1$ to $e_4$ of these heat engines.

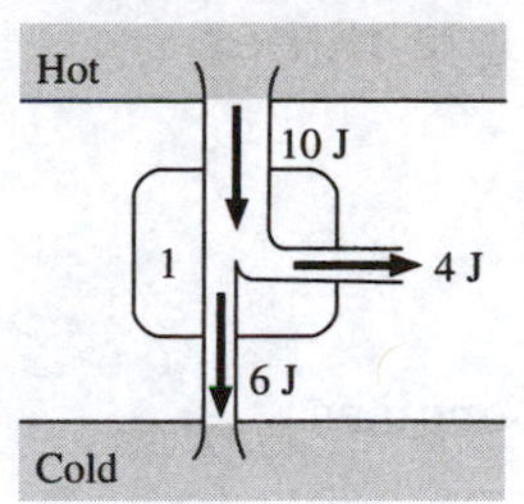


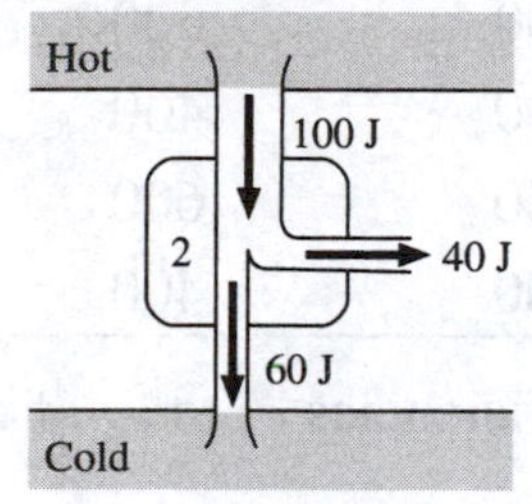


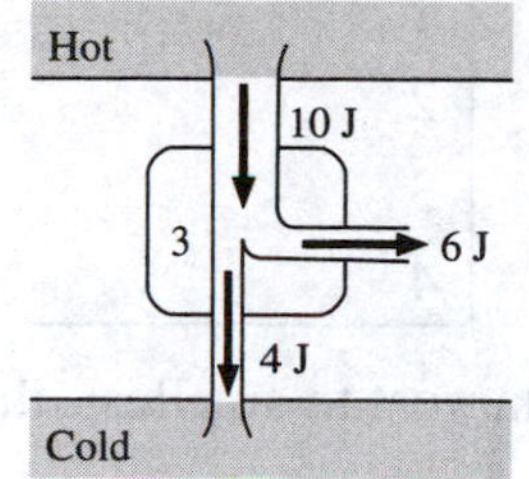


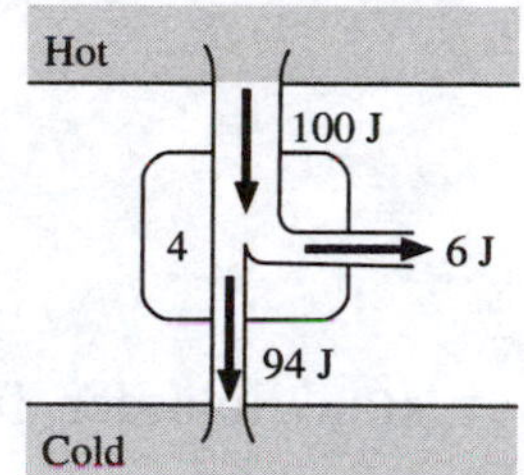


Order:

Explanation:

15. For each engine shown,
    a. Supply the missing value.
    b. Determine the efficiency.

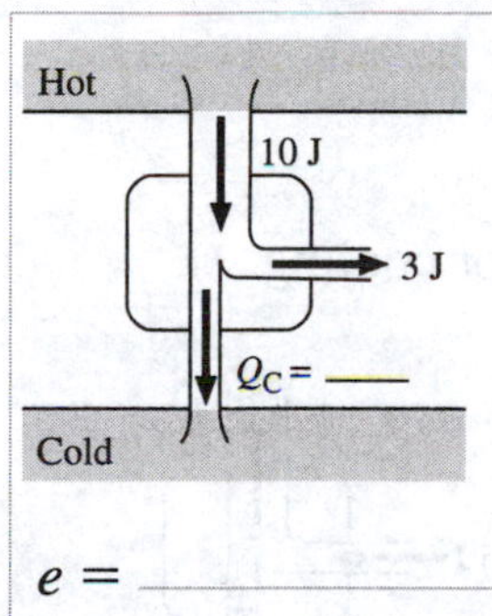


$e$ = ______

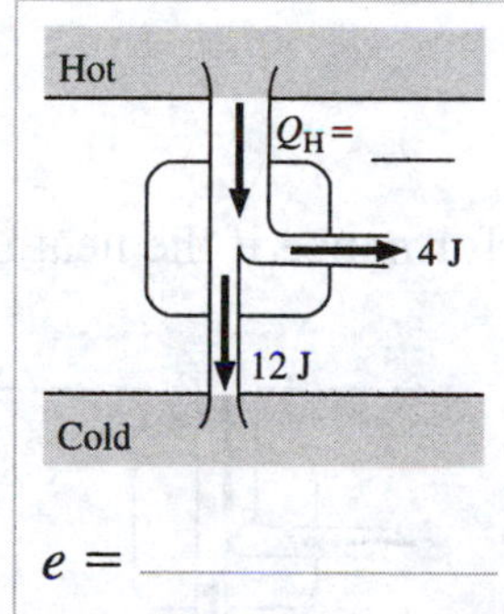


$e$ = ______

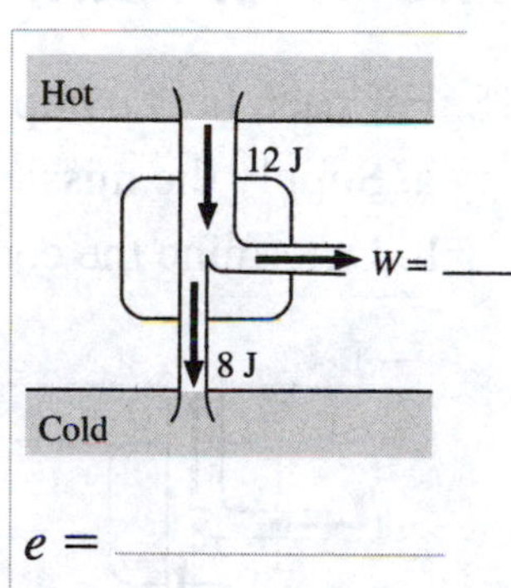


$e$ = ______

16. Efficiency is a dimensionless quantity, so why is it necessary to measure temperatures in Kelvin rather than °C or °F to determine efficiency when using the equation $e_{max} = 1 - \frac{T_C}{T_H}$?

17. Four heat engines with maximum efficiency (Carnot engines) operate with the hot and cold reservoir temperatures shown in the table.

| Engine | $T_C$ (K) | $T_H$ (K) |
|---|---|---|
| 1 | 300 | 600 |
| 2 | 200 | 400 |
| 3 | 200 | 600 |
| 4 | 300 | 400 |

Rank in order, from largest to smallest, the efficiencies $e_1$ to $e_4$ of these engines.

Order:

Explanation:

## 11.6 Heat Pumps, Refrigerators, and Air Conditioners

18. For each heat pump shown,
    a. Supply the missing value.
    b. Determine the coefficient of performance if the heat pump is used for cooling.

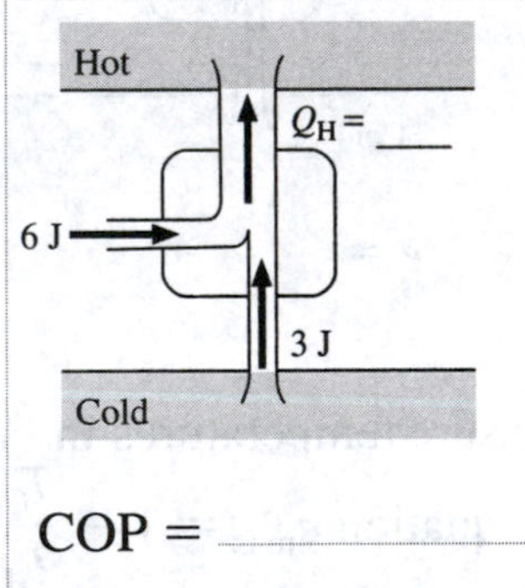


COP = ____

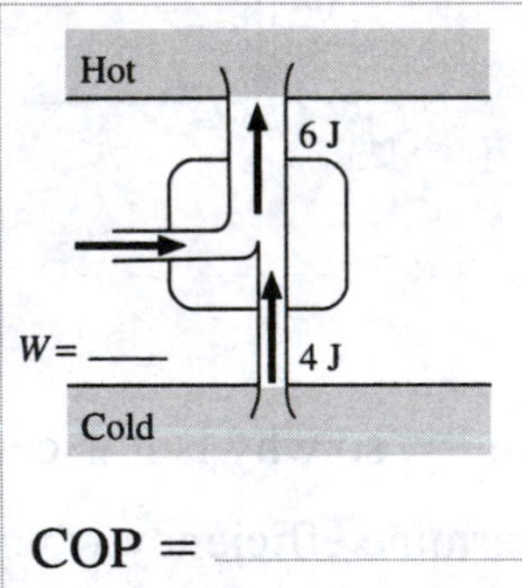


COP = ____

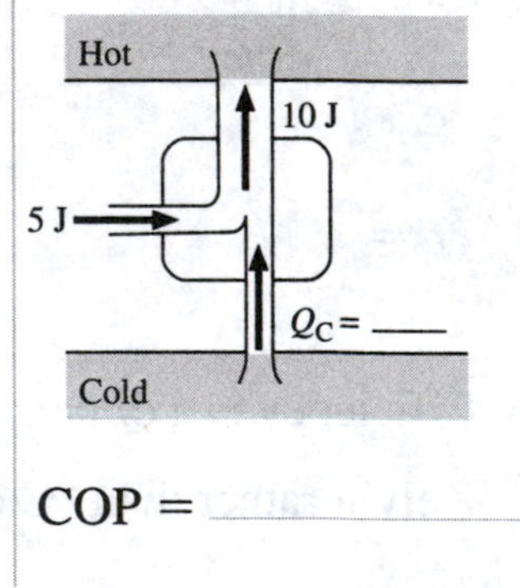


COP = ____

19. Does a refrigerator do work in order to cool the interior? Explain.

## 11.7 Entropy and the Second Law of Thermodynamics

## 11.8 Systems, Energy, and Entropy

20. Do each of the following represent a possible heat engine or heat pump? If not, what is wrong?

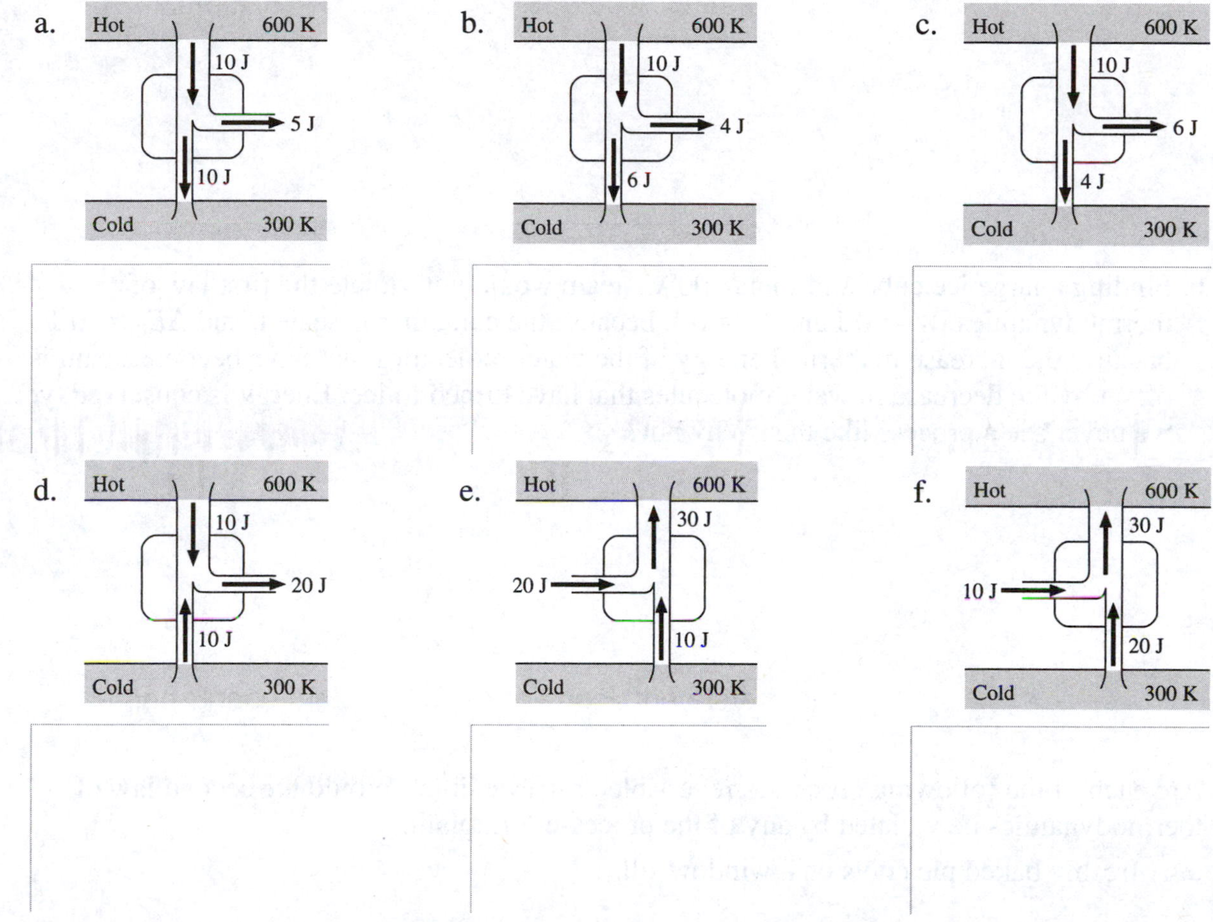


21. If you place a jar of perfume in the center of a room and remove the stopper, you will soon be able to smell the perfume throughout the room. If you wait long enough, will all the perfume molecules ever be back in the jar at the same time? Why or why not?

22. Suppose you place an ice cube in a cup of room-temperature water and then seal them in a well-insulated container. No energy can enter or leave the container.

a. If you open the container an hour later, which do you expect to find: a cup of water, slightly cooler than room temperature, or a large ice cube and some 100°C steam?

b. Finding a large ice cube and some 100°C steam would not violate the first law of thermodynamics. $W = 0$ J and $Q = 0$ J, because the container is sealed, and $\Delta E_{th} = 0$ J because the increase in thermal energy of the water molecules that have become steam is offset by the decrease in water molecules that have turned to ice. Energy is conserved, yet we never see a process like this. Why not?

23. Are each of the following processes reversible or irreversible? Would the second law of thermodynamics be violated by any of the processes? Explain.

a. A freshly baked pie cools on a window sill.

b. A neatly raked pile of leaves is scattered by the wind.

c. The wind gathers up the fallen leaves in a yard and leaves them in a neat pile.

# 12 Thermal Properties of Matter

## 12.1 The Atomic Model of Matter

1. The block shown is made up of eight identical cubes, set $2 \times 2 \times 2$.

   a. In the space at right, draw a block that has twice the total volume but still made out of cubes of the same size.

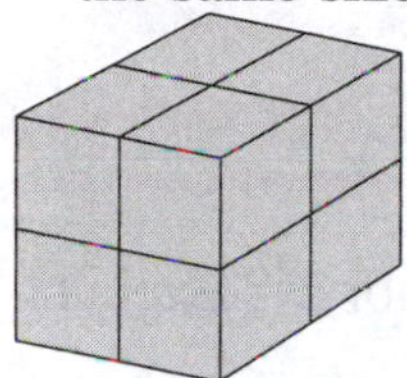

   b. How many cubes are needed to construct the larger block? ______________

   c. How many cubes are needed to construct a solid block that is twice as large in each dimension ($4 \times 4 \times 4$)?

   b. By what factor is the volume of the block in part c greater than the volume of the original $2 \times 2 \times 2$ block? ______________

2. a. Solids and liquids resist being compressed. They are not totally incompressible, but it takes large forces to compress them even slightly. If it is true that matter consists of atoms, what can you infer about the microscopic nature of solids and liquids from their incompressibility?

   b. Solids also resist being pulled apart. You can break a metal or glass rod by pulling the ends in opposite directions, but it takes a large force to do so. What can you infer from this observation about the properties of atoms?

## 12.2 The Atomic Model of an Ideal Gas

3. Gases, in contrast with solids and liquids, are very compressible. What can you infer from this observation about the microscopic nature of gases?

4. If you double the temperature of a gas,

a. Does the root-mean-square speed of the atoms increase by a factor of $(2)^{1/2}$, 2, or $2^2$? Explain.

b. Does the average kinetic energy of the atoms increase by a factor of $(2)^{1/2}$, 2, or $2^2$? Explain.

5. Suppose you could suddenly increase the speed of every atom in a gas by a factor of 2.

a. Would the rms speed of the atoms increase by a factor of $(2)^{1/2}$, 2, or $2^2$? Explain.

b. Would the thermal energy of the gas increase by a factor of $(2)^{1/2}$, 2, or $2^2$? Explain.

c. Would the temperature of the gas increase by a factor of $(2)^{1/2}$, 2, or $2^2$? Explain.

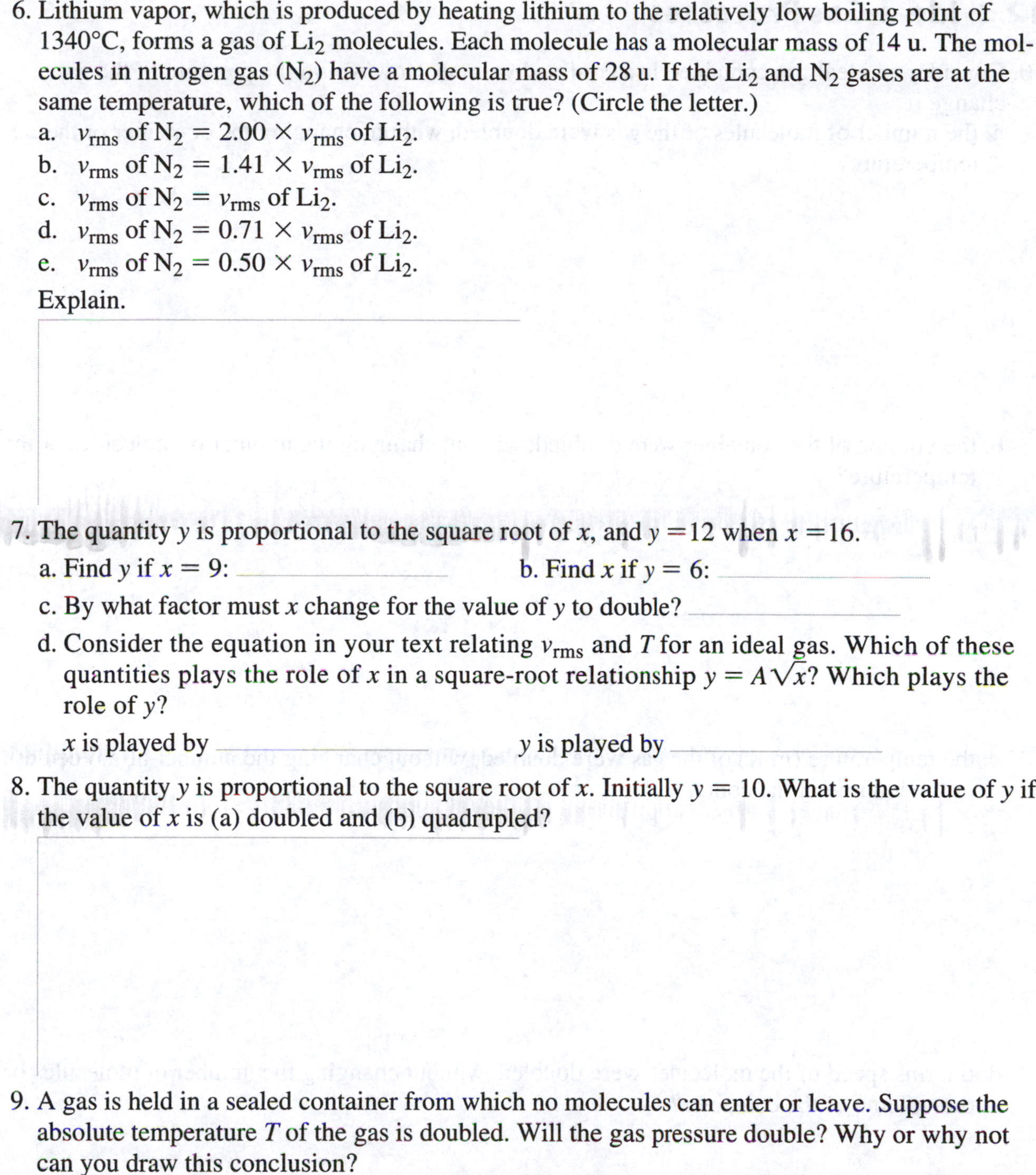

6. Lithium vapor, which is produced by heating lithium to the relatively low boiling point of 1340°C, forms a gas of $Li_2$ molecules. Each molecule has a molecular mass of 14 u. The molecules in nitrogen gas ($N_2$) have a molecular mass of 28 u. If the $Li_2$ and $N_2$ gases are at the same temperature, which of the following is true? (Circle the letter.)
   a. $v_{rms}$ of $N_2 = 2.00 \times v_{rms}$ of $Li_2$.
   b. $v_{rms}$ of $N_2 = 1.41 \times v_{rms}$ of $Li_2$.
   c. $v_{rms}$ of $N_2 = v_{rms}$ of $Li_2$.
   d. $v_{rms}$ of $N_2 = 0.71 \times v_{rms}$ of $Li_2$.
   e. $v_{rms}$ of $N_2 = 0.50 \times v_{rms}$ of $Li_2$.
   Explain.

7. The quantity $y$ is proportional to the square root of $x$, and $y = 12$ when $x = 16$.
   a. Find $y$ if $x = 9$: ________ b. Find $x$ if $y = 6$: ________
   c. By what factor must $x$ change for the value of $y$ to double? ________
   d. Consider the equation in your text relating $v_{rms}$ and $T$ for an ideal gas. Which of these quantities plays the role of $x$ in a square-root relationship $y = A\sqrt{x}$? Which plays the role of $y$?
   $x$ is played by ________ $y$ is played by ________

8. The quantity $y$ is proportional to the square root of $x$. Initially $y = 10$. What is the value of $y$ if the value of $x$ is (a) doubled and (b) quadrupled?

9. A gas is held in a sealed container from which no molecules can enter or leave. Suppose the absolute temperature $T$ of the gas is doubled. Will the gas pressure double? Why or why not can you draw this conclusion?

## 12.3 Ideal-Gas Processes

10. Consider an ideal gas contained in a confined volume. How would the pressure of the gas change if
    a. the number of molecules of the gas were doubled, without changing the container or the temperature?

    b. the volume of the container were doubled, without changing the number of molecules or the temperature?

    c. the temperature (in K) of the gas were doubled, without changing the number of molecules or the volume of the container?

    d. the rms speed of the molecules were doubled, without changing the number of molecules or the volume of the container?

11. The graphs below use a dot to show the initial state of a gas. Draw a $pV$ diagram showing the following processes:
    a. A constant-volume process that doubles the pressure.
    b. An isobaric process that doubles the temperature.
    c. An isothermal process that halves the volume.

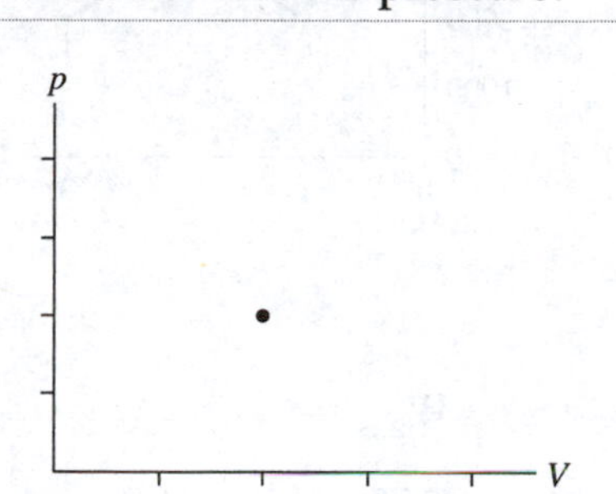


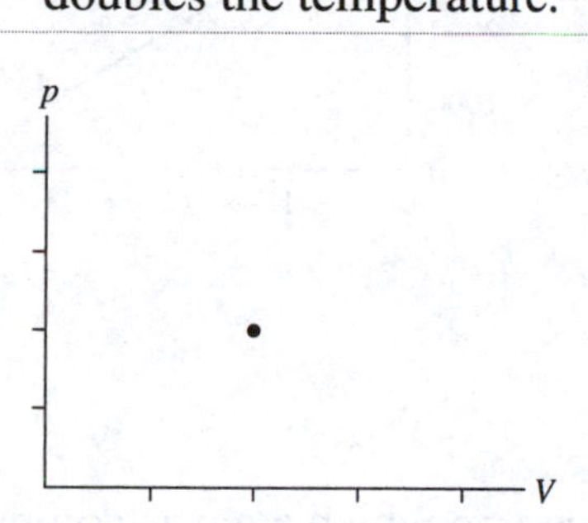


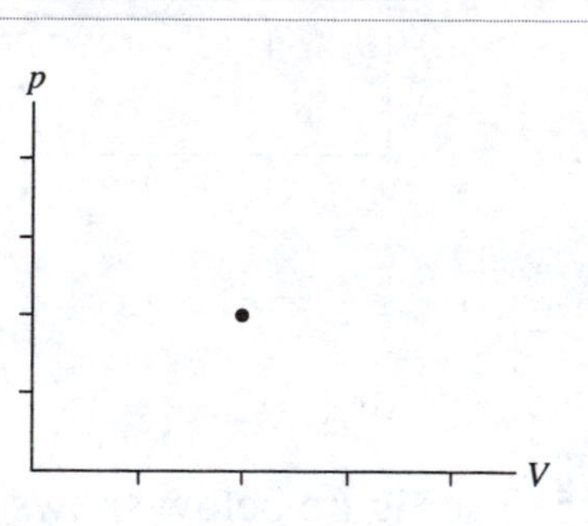


12. Interpret the $pV$ diagrams shown below by
    a. Naming the process.
    b. Stating the *factors* by which $p$, $V$, and $T$ change. (A fixed quantity changes by a factor of 1.)

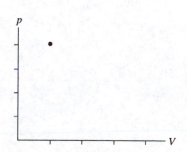

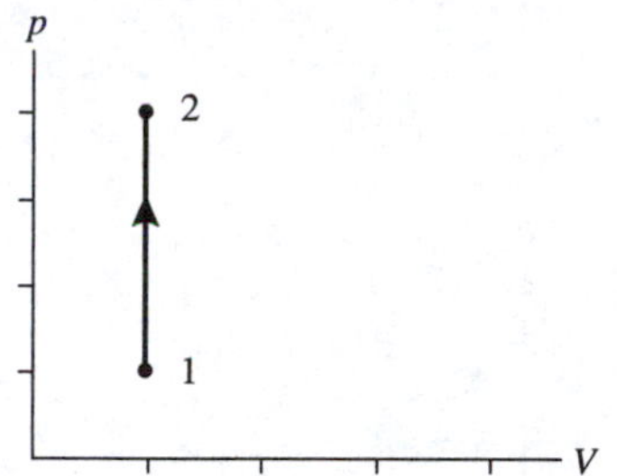


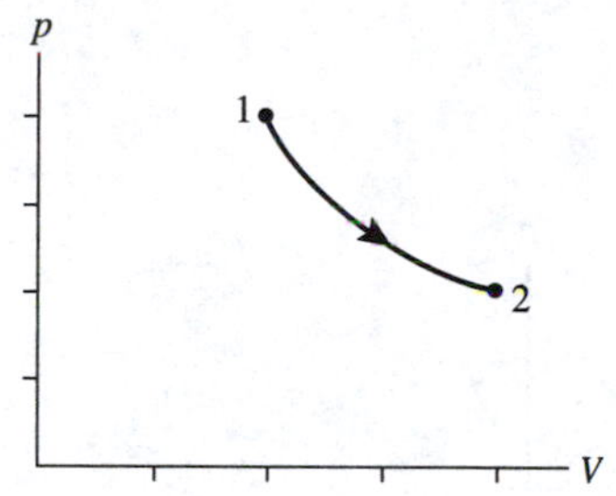


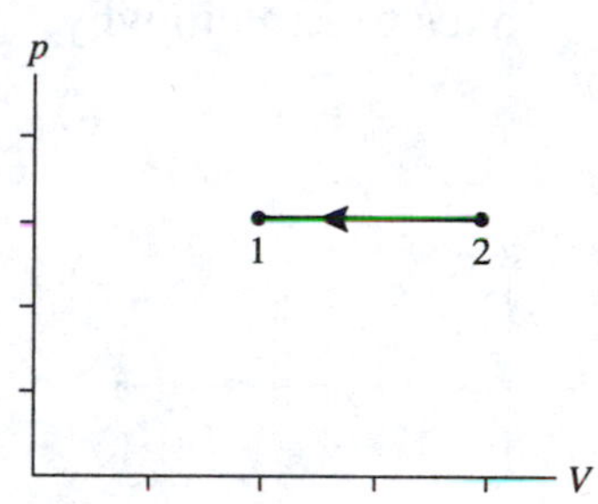


Process ______________
$p$ changes by ______________
$V$ changes by ______________
$T$ changes by ______________

Process ______________
$p$ changes by ______________
$V$ changes by ______________
$T$ changes by ______________

Process ______________
$p$ changes by ______________
$V$ changes by ______________
$T$ changes by ______________

13. Starting from the initial state shown, draw a $pV$ diagram for the three-step process:
    i. A constant-volume process that halves the temperature.
    ii. An isothermal process that halves the pressure, then
    iii. An isobaric process that doubles the volume.

    Label each of the stages on your diagram.

14. How much work is done by the gas in each of the following processes?

a. 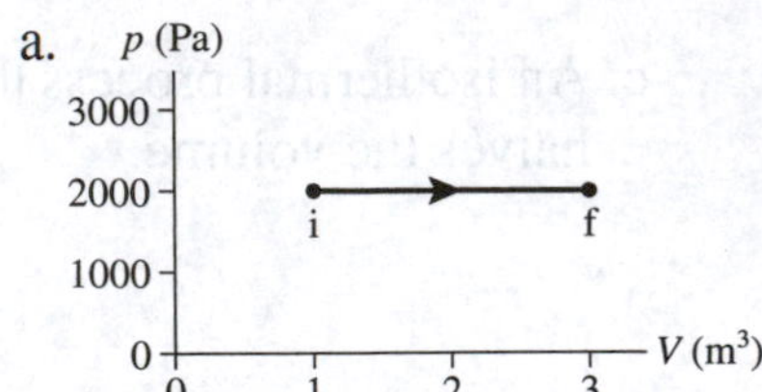


b. 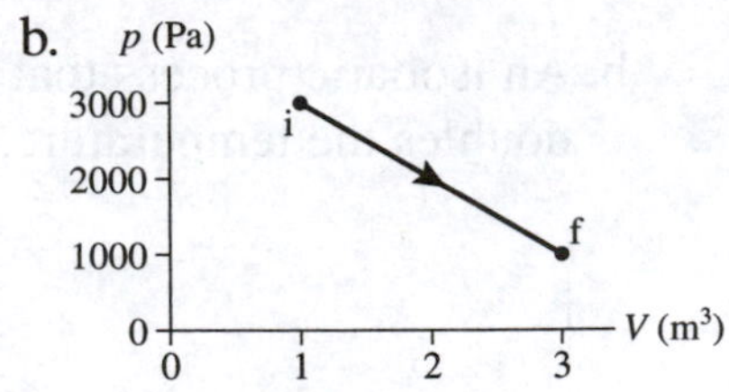


c. 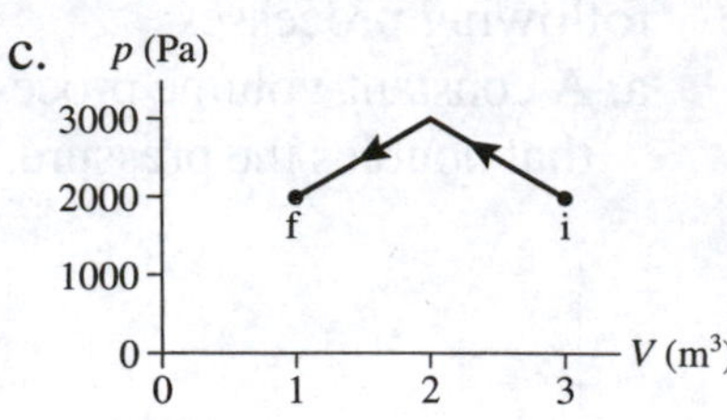


$W_{gas}$ = ______________ $W_{gas}$ = ______________ $W_{gas}$ = ______________

15. The figure below shows a process in which a gas is compressed from 300 cm$^3$ to 100 cm$^3$.

a. Use the middle set of axes to draw the $pV$ diagram of a process that starts from initial state i, compresses the gas to 100 cm$^3$, and does the same amount of work on the gas as the process on the left.

b. Is there a constant-volume process that does the same amount of work on the gas as the process on the left? If so, show it on the axes on the right. If not, use the blank space of the axes to explain why.

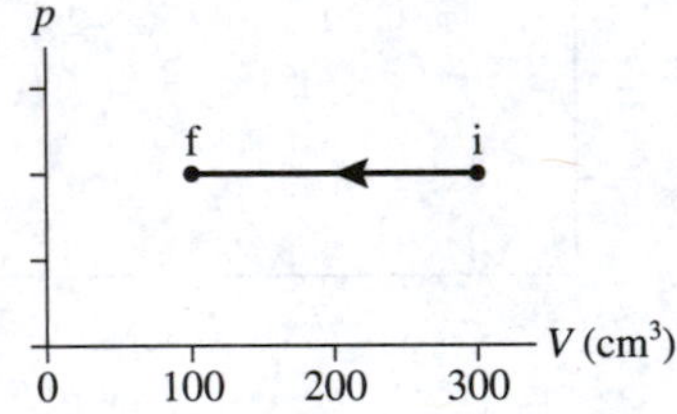


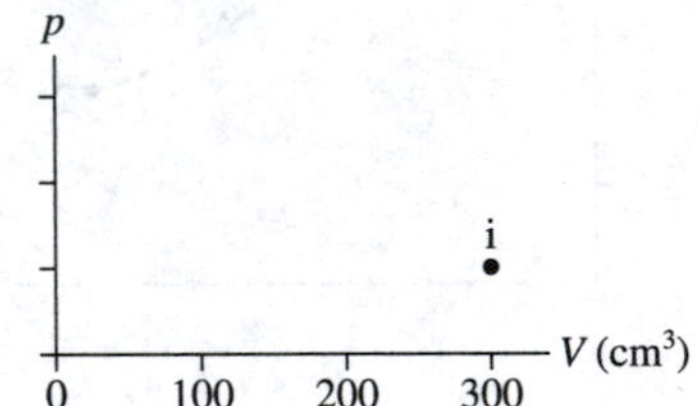


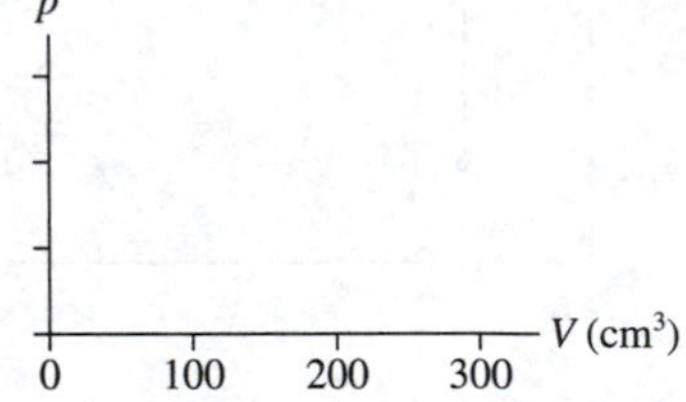


16. The figure shows a process in which work is done to compress a gas.

a. Draw and label a process A that starts and ends at the same points but does more work on the gas.

b. Draw and label a process B that starts and ends at the same points but does less work on the gas.

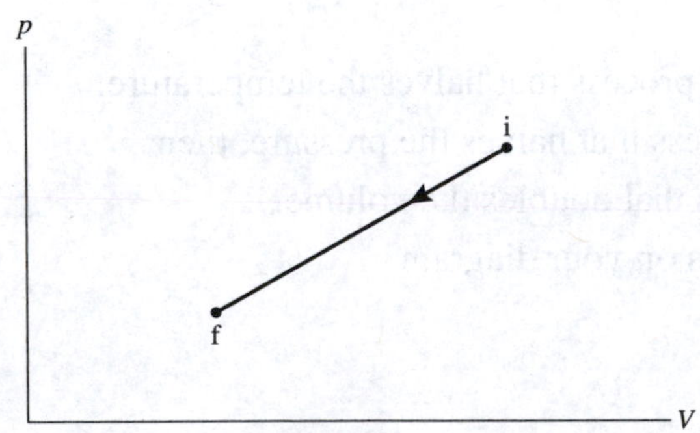

17. Starting from the point shown, draw a $pV$ diagram for the following processes.

a. An isobaric process in which work is done *by* the system.

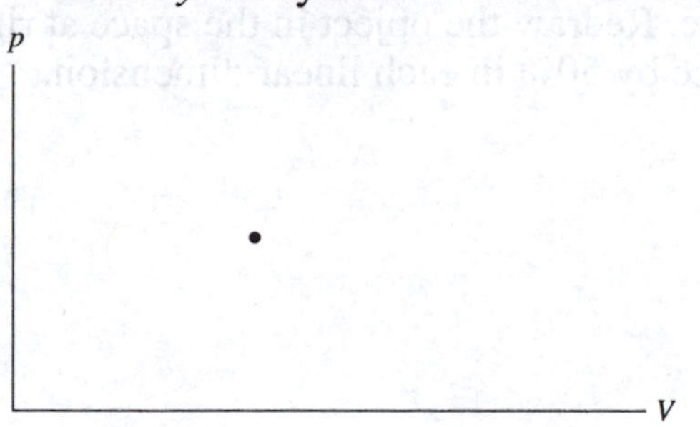


b. An adiabatic process in which work is done *on* the system.

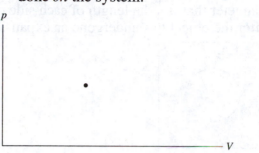

c. An isothermal process in which heat is *added to* the system.

d. A constant-volume process in which heat is *removed from* the system.

p

V

## You Write the Problem!

**Exercises 18–20:** You are given the equation that is used to solve a problem. For each of these:

a. Write a *realistic* physics problem for which this is the correct equation. Look at worked examples and end-of-chapter problems in the textbook to see what realistic physics problems are like.
b. Finish the solution of the problem.

18. $p_f = \dfrac{300\ \text{cm}^3}{100\ \text{cm}^3} \times 1 \times 2\ \text{atm}$

19. $(T_f + 273)\ \text{K} = \dfrac{200\ \text{kPa}}{500\ \text{kPa}} \times 1 \times (400 + 273)\ \text{K}$

20. $V_f = \dfrac{(400 + 273)\ \text{K}}{(50 + 273)\ \text{K}} \times 1 \times 200\ \text{cm}^3$

## 12.4 Thermal Expansion

21. The figure shows a thin square solid object with a circular hole in the center. The hole has a diameter that is $\frac{1}{2}$ the length of each side of the square. Redraw the object in the space at right after the object has undergone an expansion of its size by 50% in each linear dimension.

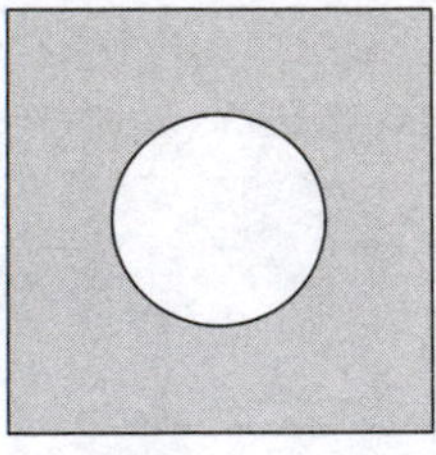

22. a. Solids are characterized by both a coefficient of linear expansion and a coefficient of volume expansion. Examine Table 12.3 and describe the relationship between these two coefficients.

b. Liquids, in contrast with solids, have only a coefficient of volume expansion. Why don't liquids have a coefficient of linear expansion?

23. Containers A and B hold equal volumes of water at the same temperature. The temperatures of both containers are increased by the same amount. Does the height of the water in A *go up* by a larger amount, a smaller amount, or the same amount as the water goes up in B? Explain.

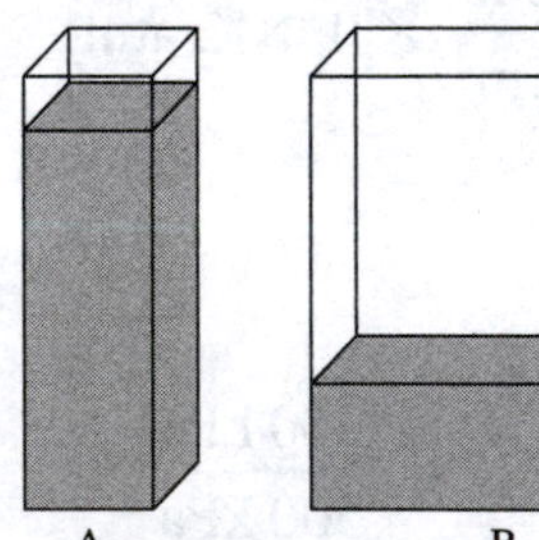

## 12.5 Specific Heat and Heat of Transformation

24. Explain in your own words the meaning of the statement that the specific heat of a liquid is 4000 J/kg·K.

25. Object A has a larger specific heat than object B. Which object's temperature will increase the most if both absorb the same amount of heat energy? Explain.

26. 100 g of water at 10°C and 100 g of water at 90°C each absorb the same amount of heat energy. Which sample's temperature increases the most? Or do they increase by the same amount? Explain. You can assume that the amount of heat absorbed is not enough to boil the 90° sample.

27. The heats of fusion of water and nitrogen are 333 kJ/kg and 26 kJ/kg, respectively. The heats of vaporization of water and nitrogen are 2260 kJ/kg and 1990 kJ/kg, respectively. Suppose 100 g of liquid water at 0°C is poured into a flask of liquid nitrogen at its –196°C boiling point. The water will freeze, and in doing so it will boil off some nitrogen. Will the process of turning the 0°C water into solid ice at 0°C boil off less than 100 g, more than 100 g, or exactly 100 g of nitrogen? Explain.

## 12.6 Calorimetry

28. A beaker of water at 80.0°C is placed in the center of a large, well-insulated room whose air temperature is 20.0°C. Is the final temperature of the water:

    i. 20.0°C.
    ii. Slightly above 20.0°C.
    iii. 50.0°C.
    iv. Slightly less than 80.0°C.
    v. 80.0°C.

    Explain.

29. You have two 100 g cubes A and B, made of different materials. Cube A has a larger specific heat than cube B. Cube A, initially at 0°C, is placed in good thermal contact with cube B, initially at 200°C. The cubes are inside a well-insulated container where they don't interact with their surroundings. Is their final temperature greater than, less than, or equal to 100°C? Explain.

## You Write the Problem!

**Exercise 30:** You are given the equation that is used to solve a problem.

a. Write a *realistic* physics problem for which this is the correct equation. Look at worked examples and end-of-chapter problems in the textbook to see what realistic physics problems are like.
b. Finish the solution of the problem.

30. $(2.72\ \text{kg})(140\ \text{J/kg}\cdot\text{K})(90°\text{C} - 15°\text{C}) + (0.50\ \text{kg})(449\ \text{J/kg}\cdot\text{K})(90°\text{C} - T_\text{i}) = 0$

## 12.7 Specific Heats of Gases

31. Explain *why* the molar specific heat at constant pressure is larger than the molar specific heat at constant volume.

32. Explain why the molar specific heat at constant volume of diatomic gases is larger than for monatomic gases.

33. You need to raise the temperature of a gas by 10°C. To use the least amount of heat energy, should you heat the gas at constant pressure or at constant volume? Explain.

34. The $pV$ diagram shows two processes, A and B, that take an ideal gas from initial state i to final state f.

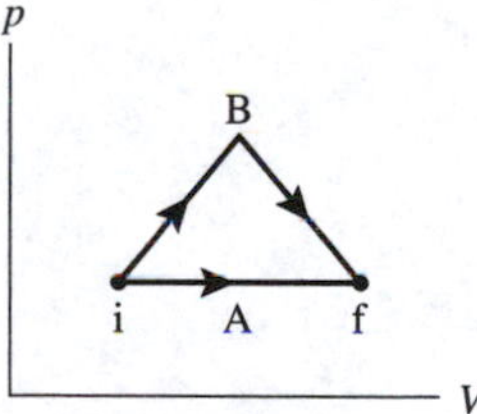


a. Is more work done by the gas in process A or in process B? Or is $W_{gas}$ the same for both? Explain.

b. According to the first law of thermodynamics, heat energy is required for both processes. Is more heat energy required for process A or for process B? Or is $Q$ the same for both? Explain.

## 12.8 Heat Transfer

35. Heat flows through a cylindrical copper bar at the rate of 100 W. A second cylindrical copper bar between the same two temperatures is twice as long and has twice the diameter as the first bar. What is the rate of heat flow through the second bar?

36. The thermal conductivities of wood and concrete are 0.2 W/m·K and 0.8 W/m·K, respectively. Suppose your house has 1-inch-thick wood siding on the outside. You would like to replace the wood siding with a concrete exterior wall. What thickness must the concrete have to provide the same insulating value as the wood? Explain.

37. A titanium cube at 400 K radiates 50 W of heat. How much heat does the cube radiate if its temperature is increased to 800 K?

38. Sphere A has radius $R$ and is at temperature $T$. Sphere B has radius $R/2$ and is at temperature $2T$. Which sphere radiates more heat? Explain.

# 13 Fluids

## 13.1 Fluids and Density

1. An object has density $\rho$.

   a. Suppose each of the object's three dimensions is increased by a factor of 2 without changing the material of which the object is made. Will the density change? If so, by what factor? Explain.

   b. Suppose each of the object's three dimensions is increased by a factor of 2 without changing the object's mass. Will the density change? If so, by what factor? Explain.

2. A stone cutter cuts a section off of a marble slab that is one-eighth the weight of the original slab. Is the density of the cut section less than, more than, or the same as the density of the original slab? Explain.

3. A cylinder contains 2 g of oxygen gas. A piston is used to compress the gas. After the gas has been compressed:

a. Has the mass of the gas increased, decreased, or not changed? Explain.

b. Has the density of the gas increased, decreased, or not changed? Explain.

4. Air enclosed in a cylinder has density $\rho = 1.4 \text{ kg/m}^3$.

a. What will be the density of the air if the length of the cylinder is doubled while the radius is unchanged?

b. What will be the density of the air if the radius of the cylinder is halved while the length is unchanged?

## 13.2 Pressure

## 13.3 Measuring and Using Pressure

5. Rank in order, from largest to smallest, the pressures $p_A$ to $p_D$ in containers A through D at the depths indicated by the dashed line.

Order:

Explanation:

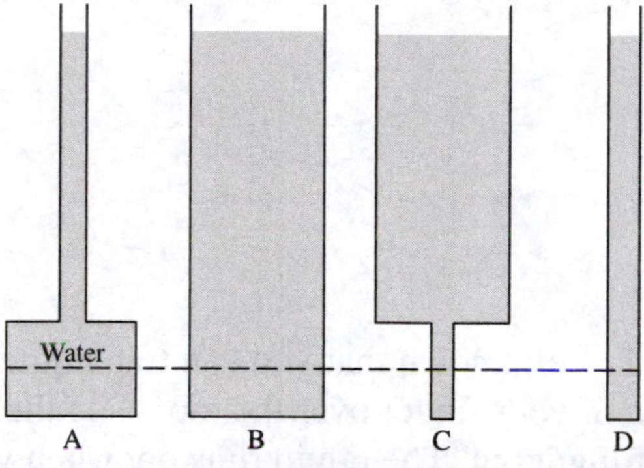


6. A and B are rectangular tanks full of water. They have equal depths, equal thicknesses (the dimension into the page) but different widths.
   a. Compare the forces the water exerts on the bottoms of the tanks. Is $F_A$ larger than, smaller than, or equal to $F_B$? Explain.

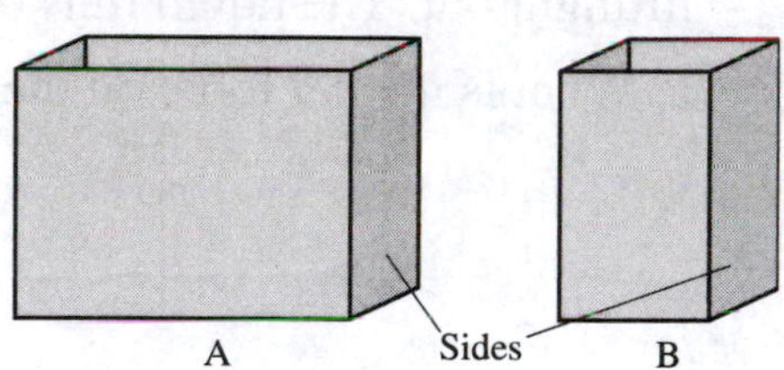


   b. Compare the forces the water exerts on the sides of the tanks. Is $F_A$ larger than, smaller than, or equal to $F_B$? Explain.

7. Is $p_A$ larger than, smaller than, or equal to $p_B$? Explain.

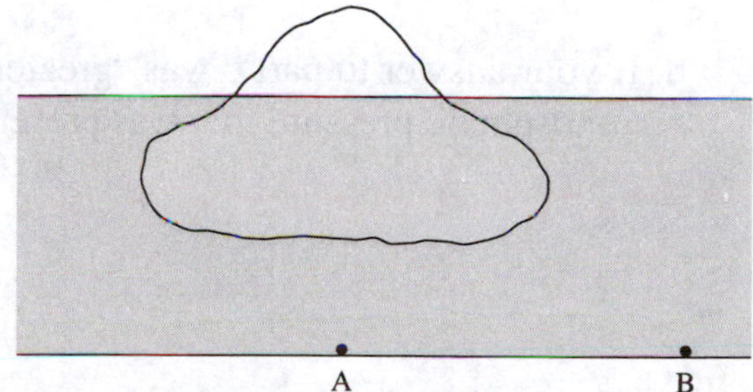

8. The figure shows a manometer like that of Figure 13.10 in the textbook, but here the height of the liquid is higher on the left side than on the right. Is this possible? If not, why not? If so, what can you say about the gas pressure in the tank?

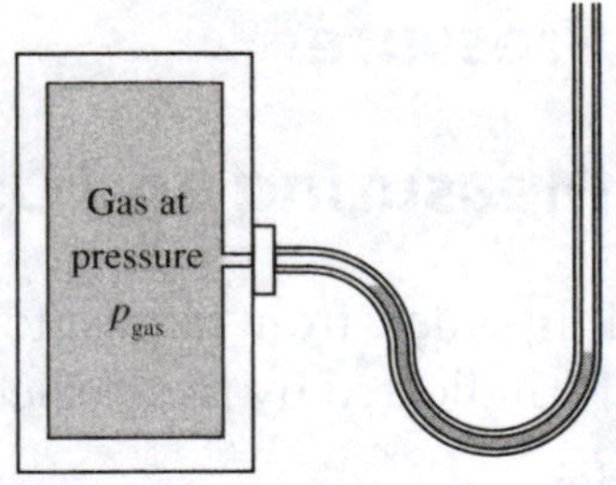


9. It is well known that you can trap liquid in a drinking straw by placing the tip of your finger over the top while the straw is in the liquid, and then lifting it out. The liquid runs out when you release your finger.

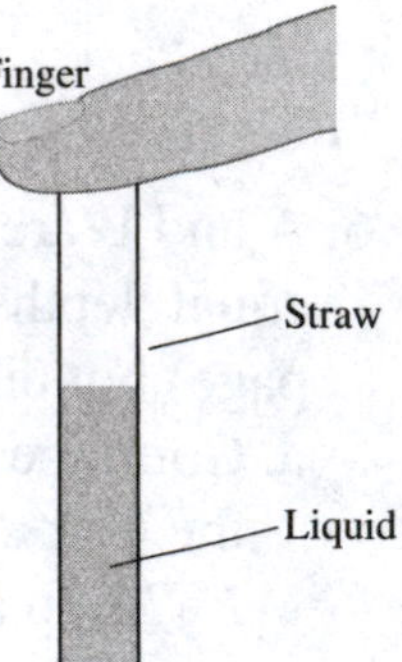


a. What is the *net* force on the cylinder of trapped liquid?

b. Three forces act on the trapped liquid. Draw and label all three on the figure.

c. Is the gas pressure inside the straw, between the liquid and your finger, greater than, less than, or equal to atmospheric pressure? Explain, basing your explanation on your answers to parts a and b.

d. If your answer to part c was "greater" or "less," how did the pressure change from the atmospheric pressure that was present when you placed your finger over the top of the straw?

## 13.4 Buoyancy

10. Three blocks of identical size, A, B, and C, are to be gently placed into a large tank of water. Block A has a density of 2 g/cm$^3$, block B has a density of 0.9 g/cm$^3$, and block C has a density of 0.5 g/cm$^3$. On the figure, draw and label each block in its final equilibrium position in the water.

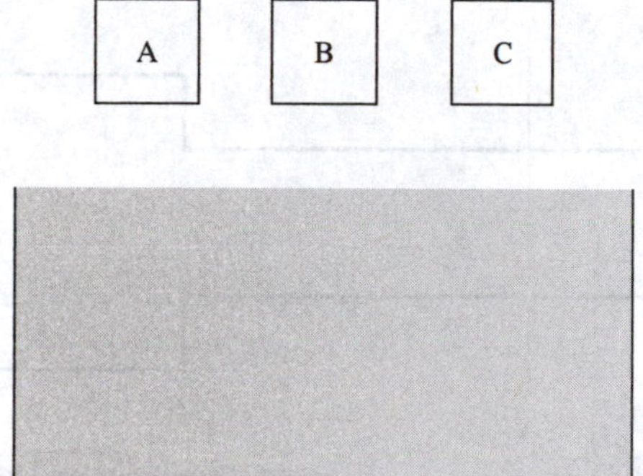


11. Rank in order, from the largest to smallest, the densities of blocks A, B, and C.

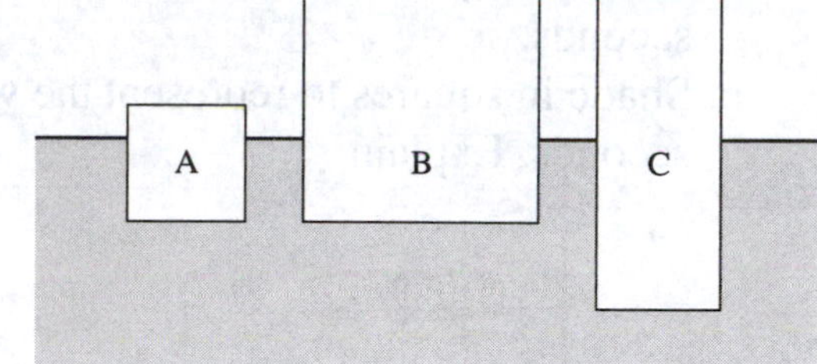


Order:

Explanation:

12. Blocks A, B, and C have the same volume. Rank in order, from largest to smallest, the sizes of the buoyant forces $F_A$, $F_B$, and $F_C$ on A, B, and C.

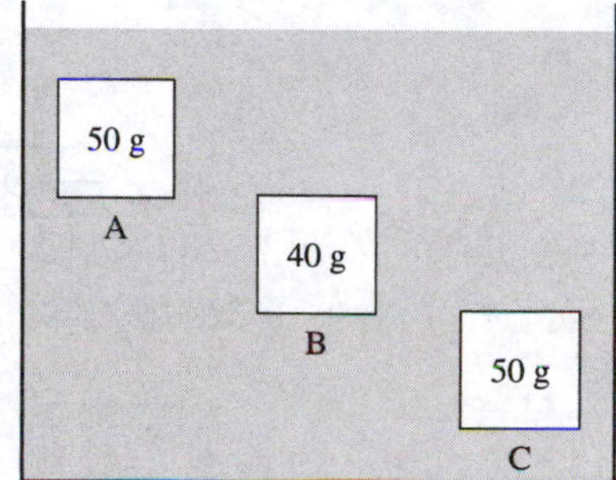


Order:

Explanation:

## 13.5 Fluids in Motion

## 13.6 Fluid Dynamics

13. A stream flows from left to right through the constant-depth channel shown below in an overhead view. A 1 m × 1 m grid has been added to facilitate measurement. The fluid's flow speed at A is 2 m/s.

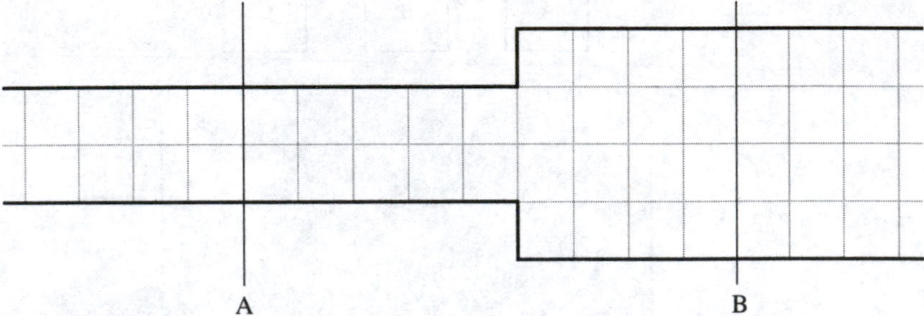


a. Shade in squares to represent the water that has flowed past point A in the last two seconds.
b. Shade in squares to represent the water that has flowed past point B in the last two seconds. Explain.

14. Liquid flows through a tube whose width varies as shown. The liquid level is shown for pipe A, which is open at the top, but not for pipes B and C. Draw an appropriate level of liquid in pipes B and C to indicate the fluid pressures at those points.

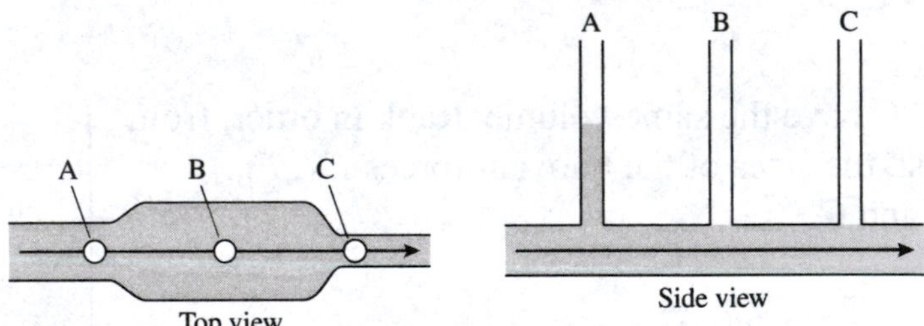

15. Gas flows through a pipe. You can't see into the pipe to know how the inner diameter changes. Rank in order, from largest to smallest, the gas speeds $v_1$ to $v_3$ at points 1, 2, and 3.

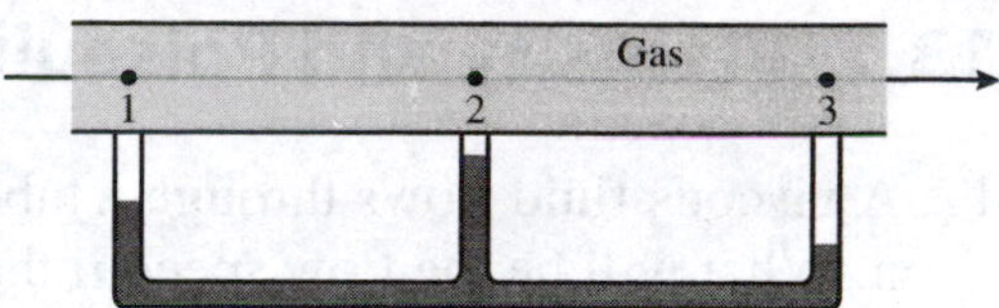


Order:

Explanation:

16. Wind blows over a house. A window on the ground floor is open. Is there an air flow through the house? If so, does the air flow in the window and out the chimney, or in the chimney and out the window? Explain.

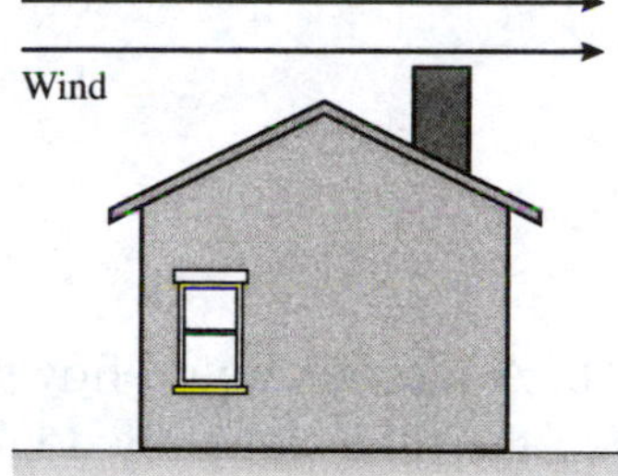

## 13.7 Viscosity and Poiseuille's Equation

17. A viscous fluid flows through a tube at an average speed of 50 cm/s.
    a. What will be the flow speed if the pressure difference between the ends of the tube is doubled but the tube's diameter is unchanged?

    b. What will be the flow speed if the tube's diameter is doubled but the pressure difference between the ends of the tube is unchanged?

18. A viscous liquid flows through a long tube. Halfway through, at the midpoint, the diameter of the tube suddenly doubles. Suppose the pressure difference between the entrance to the tube and the tube's midpoint is 800 Pa. What is the pressure difference between the midpoint of the tube and the exit?

# 14 Oscillations

## 14.1 Equilibrium and Oscillation

## 14.2 Linear Restoring Forces and SHM

1. On the axes below, sketch three cycles of the position-versus-time graph for:

   a. A particle undergoing simple harmonic motion.

   b. A particle undergoing periodic motion that is not simple harmonic motion.

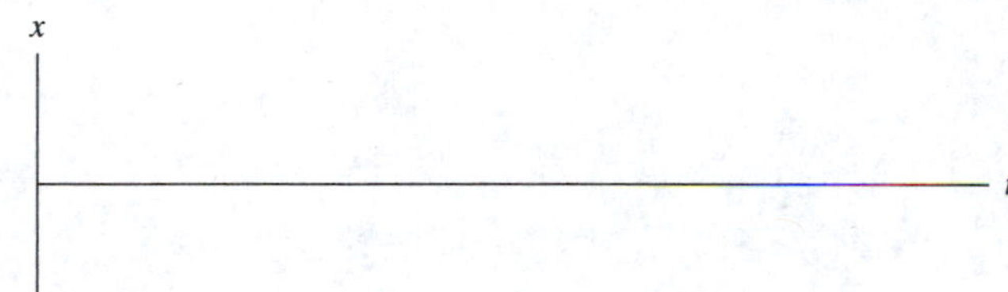


2. Consider the particle whose motion is represented by the $x$-versus-$t$ graph below.

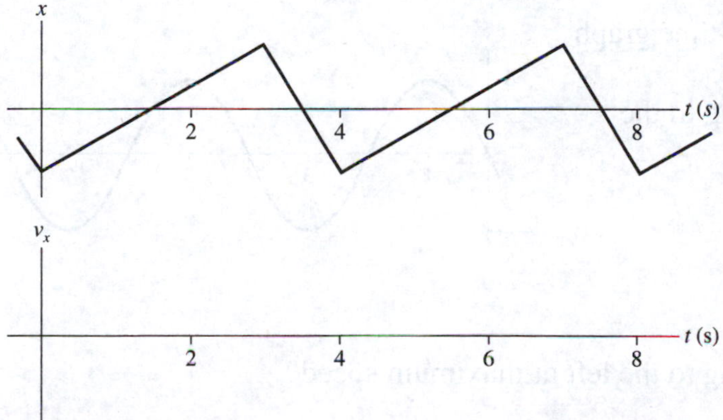


   a. Is this periodic motion? ____________ b. Is this motion SHM? ____________

   c. What is the period? ____________ d. What is the frequency? ____________

   e. You learned in Chapter 2 to relate velocity graphs to position graphs. Use that knowledge to draw the particle's velocity-versus-time graph on the axes provided.

3. Figure A shows an unstretched spring. Figure B shows a mass $m$ hanging at rest from the spring. It has stretched the spring by $L$. Figures C and D show two instants in the oscillation of mass $m$ about the equilibrium position. Draw free-body diagrams for the mass in the Figures B, C, and D.

A B C D

$k$

Unstretched spring

$L$

$m$

Equilibrium position

$m$

$m$

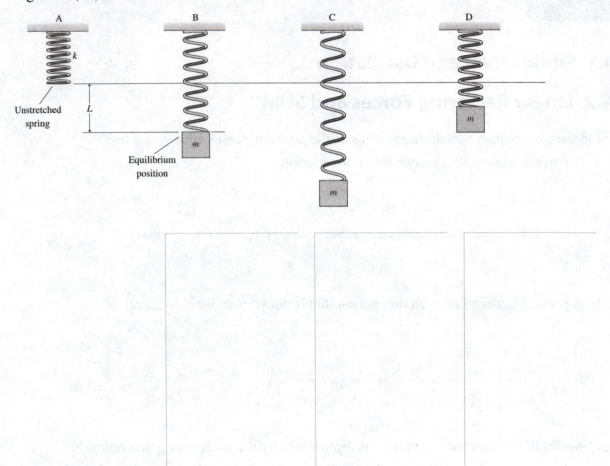

4. The figure shows the position-versus-time graph of a particle in SHM.
   a. At what times is the particle moving to the right at maximum speed?

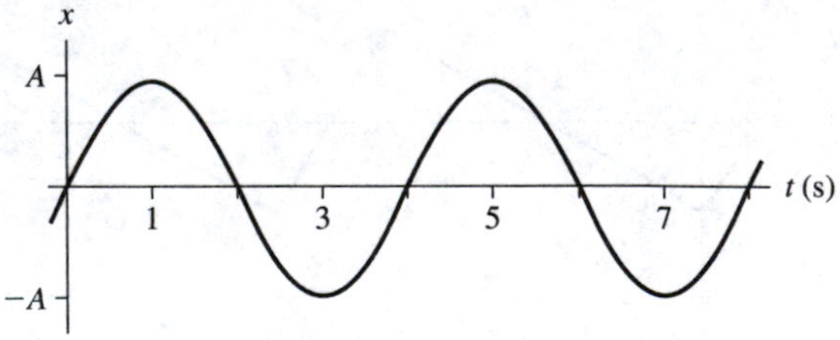


   b. At what times is the particle moving to the left at maximum speed?

   c. At what times is the particle instantaneously at rest?

## 14.3 Describing Simple Harmonic Motion

5. The graph shown is the position-versus-time graph of an oscillating particle.
   a. Draw the corresponding velocity-versus-time graph.
   b. Draw the corresponding acceleration-versus-time graph.

   Hint: Remember that velocity is the slope of the position graph, and acceleration is the slope of the velocity graph.

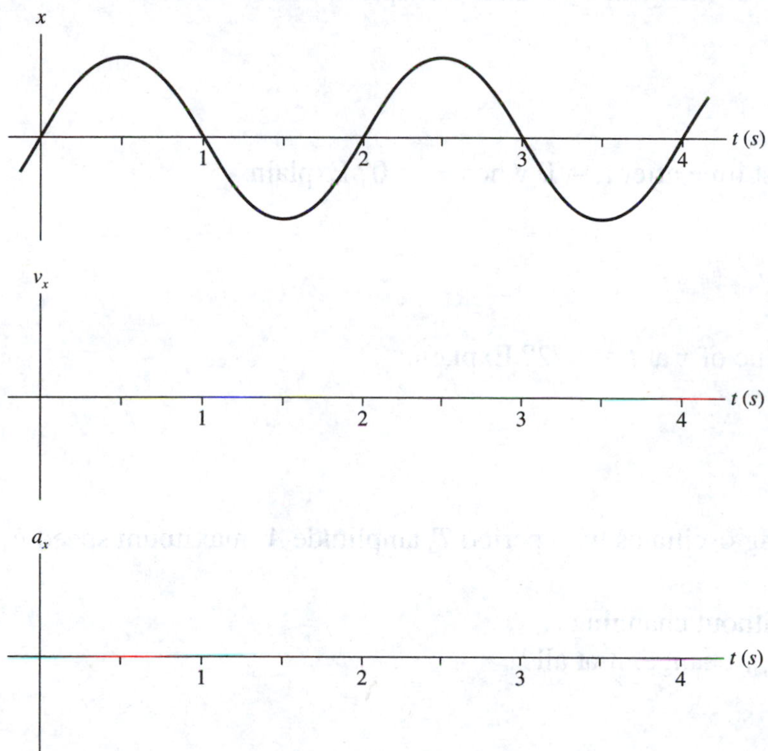


   c. At what times is the position a maximum? ________

   At those times, is the velocity a maximum, a minimum, or zero? ________

   At those times, is the acceleration a maximum, a minimum, or zero? ________

   d. At what times is the position a minimum (most negative)? ________

   At those times, is the velocity a maximum, a minimum, or zero? ________

   At those times, is the acceleration a maximum, a minimum, or zero? ________

   e. At what times is the velocity a maximum? ________

   At those times, where is the particle? ________

   f. Can you find a simple relationship between the *sign* of the position and the *sign* of the acceleration at the same instant of time? If so, what is it?

6. Consider the function $x(t) = A\cos(2\pi t/T)$.
   a. What are the units of the quantity $2\pi t/T$? ______
   b. What is the value of $x$ at $t = 0$? Explain.

   c. What are the next *two* times at which $x$ has the same value as it does at $t = 0$? Your answer will be in terms of the quantity $T$ and, perhaps, constants such as $\pi$.

   d. What is the first time after $t = 0$ when $x = 0$? Explain.

   e. What is the value of $x$ at $t = T/2$? Explain.

7. A mass on a spring oscillates with period $T$, amplitude $A$, maximum speed $v_{max}$, and maximum acceleration $a_{max}$.
   a. If $T$ doubles without changing $A$,
      i. how does $v_{max}$ change, if at all?

      ii. how does $a_{max}$ change, if at all?

   b. If $A$ doubles without changing $T$,
      i. how does $v_{max}$ change, if at all?

      ii. how does $a_{max}$ change, if at all?

## 14.4 Energy in Simple Harmonic Motion

8. The figure shows a graph of the potential energy of a block oscillating on a spring. The horizontal line represents the block's total energy $E$.

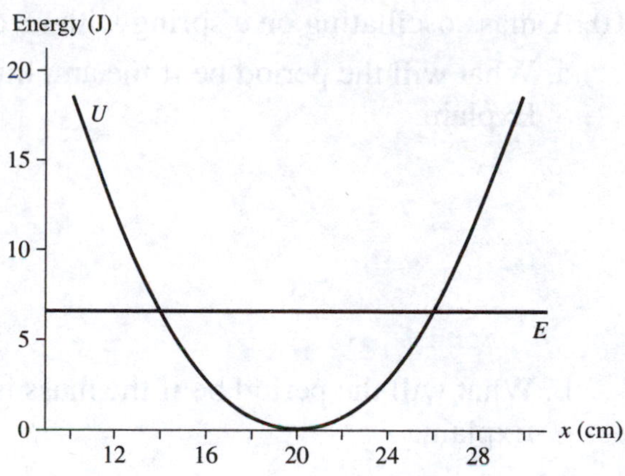


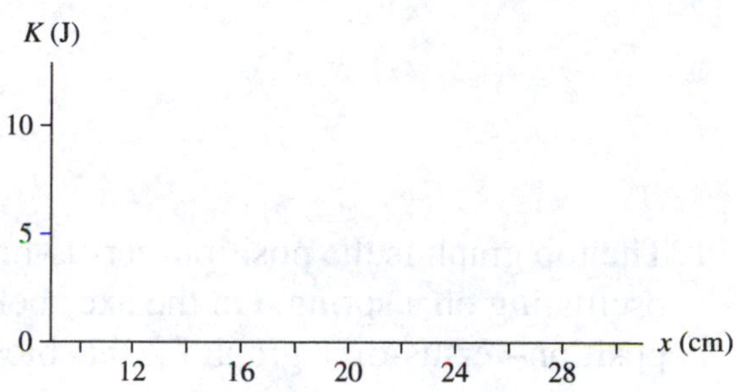


a. What is the spring's equilibrium length?

b. Where are the turning points of the motion? Explain how you identify them.

c. What is the block's maximum kinetic energy?

d. Draw a graph of the block's kinetic energy as a function of position.

e. What will be the turning points if the block's total energy is doubled?

9. Equation 14.22 in the textbook states that $\frac{1}{2}m(v_{max})^2 = \frac{1}{2}kA^2$. What does this mean? Write a couple of sentences explaining how to interpret this equation.

10. A mass oscillating on a spring with an amplitude of 5.0 cm has a period of 2.0 s.

a. What will the period be if the amplitude is doubled to 10.0 cm without changing the mass? Explain.

b. What will the period be if the mass is doubled without changing the 5.0 cm amplitude? Explain.

11. The top graph is the position-versus-time graph for a mass oscillating on a spring. On the axes below, sketch the position-versus-time graph for this block for the following situations:

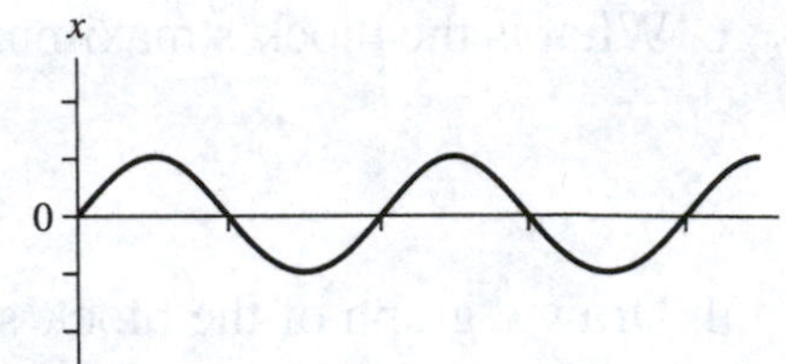


**Note:** The changes described in each part refer back to the original oscillation, not to the oscillation of the previous part of the question. Assume that all other parameters remain constant. Use the same horizontal and vertical scales as the original oscillation graph.

a. The amplitude and the frequency are both doubled.

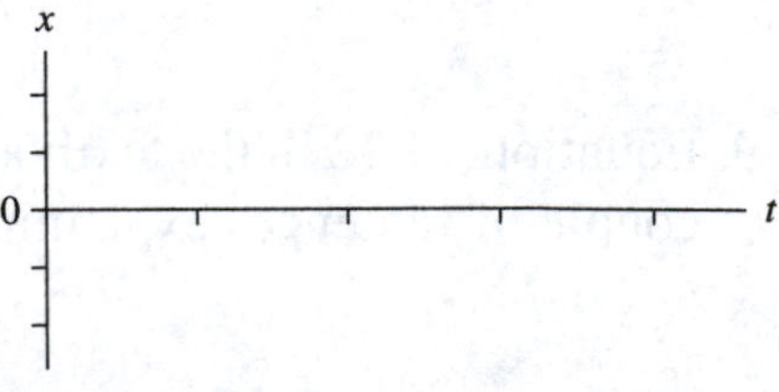


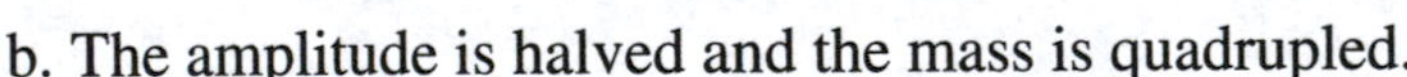

b. The amplitude is halved and the mass is quadrupled.

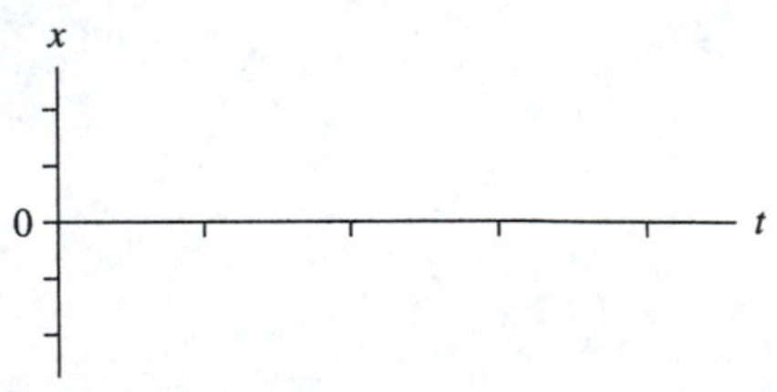


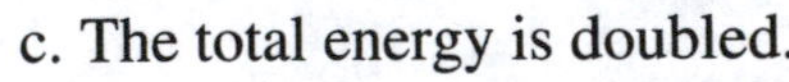

c. The total energy is doubled.

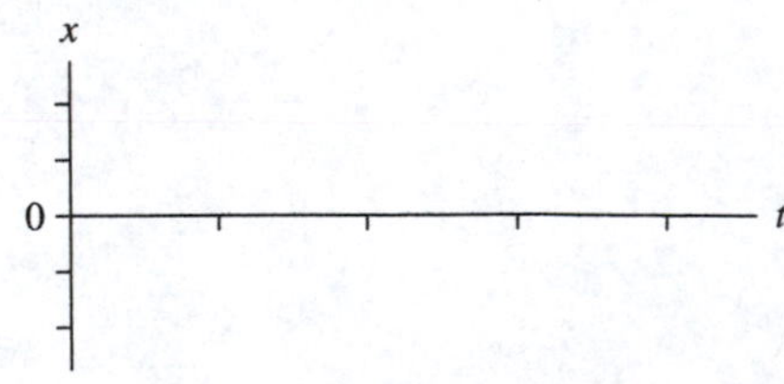

## 14.5 Pendulum Motion

12. The graph shows the displacement $s$ versus time for an oscillating pendulum.

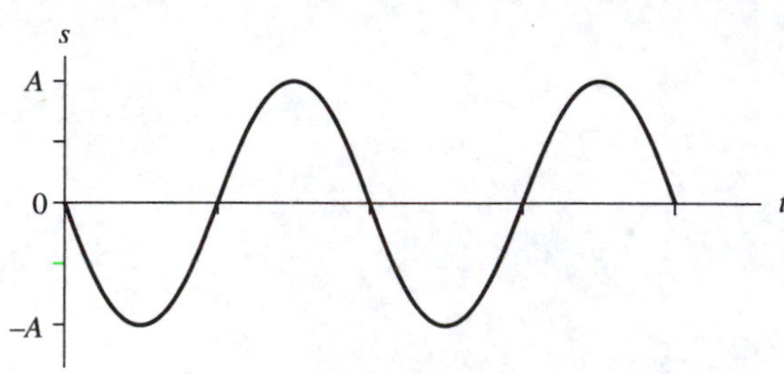


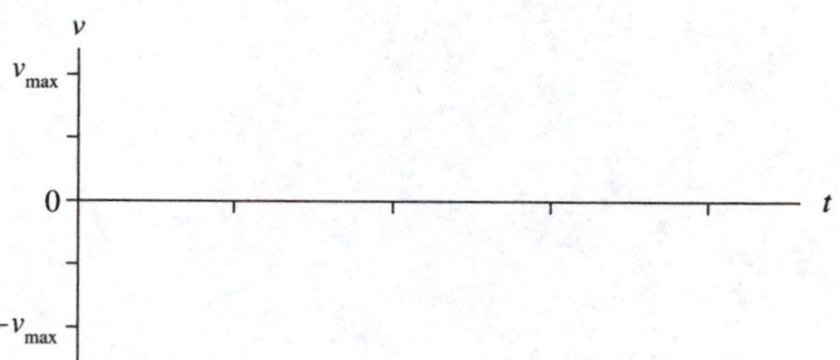


a. Draw the pendulum's velocity-versus-time graph.

b. In the space at the right, draw a *picture* of the pendulum that shows (and labels!)
   - The extremes of its motion.
   - Its position at $t = 0$ s.
   - Its direction of motion (using an arrow) at $t = 0$ s.

13. A pendulum on planet X, where the value of $g$ is unknown, oscillates with a period of 2 seconds. What is the period of this pendulum if:

a. Its mass is increased by a factor of 4?

**Note:** You do not know the values of $m$, $L$, or $g$, so do not assume any specific values.

b. Its length is increased by a factor of 4?

c. Its oscillation amplitude is increased by a factor of 4?

## 14.6 Damped Oscillations

14. For a damped oscillation, is the time constant $\tau$ greater than or less than the time in which the oscillation amplitude decays to half of its initial value? Explain.

15. The amplitude of a damped oscillation decays to one-half of its initial value in 4.0 s. How much *additional* time will it take until the amplitude is one-quarter of its initial value? Explain.

16. If the time constant $\tau$ of an oscillator is decreased, do the oscillations die away more quickly or less quickly? Explain.

17. The figure below shows the decreasing amplitude of a damped oscillator. (The oscillations are occurring rapidly and are not shown; this shows only their amplitude.) On the same axes, draw the amplitude if (a) the time constant is doubled and (b) the time constant is halved. Label your two curves "doubled" and "halved."

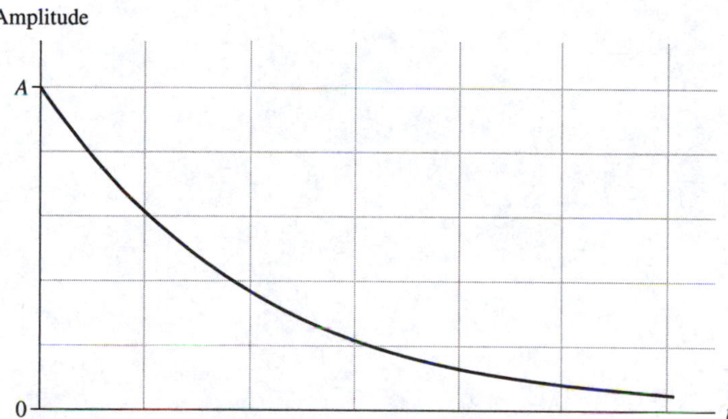

## 14.7 Driven Oscillations and Resonance

18. A car drives along a bumpy road on which the bumps are equally spaced. At a speed of 20 mph, the frequency of hitting bumps is equal to the natural frequency of the car bouncing on its springs.

    a. Draw a graph of the car's vertical bouncing amplitude as a function of its speed if the car has new shock absorbers (large damping coefficient).

    b. Draw a graph of the car's vertical bouncing amplitude as a function of its speed if the car has worn out shock absorbers (small damping coefficient).

    Draw both graphs on the same axes, and label them as to which is which.

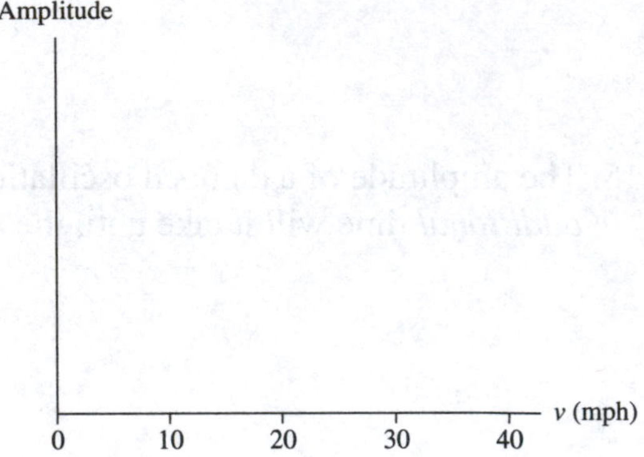


## You Write the Problem!

**Exercises 19–20:** You are given the equation that is used to solve a problem. For each of these:

a. Write a *realistic* physics problem for which this is the correct equation. Look at worked examples and end-of-chapter problems in the textbook to see what realistic physics problems are like. Be sure that the problem you write, and the answer you ask for, is consistent with the information given in the equation.

b. Finish the solution of the problem.

19. $1.7\text{ m/s} = \dfrac{2\pi A}{0.75\text{ s}}$

20. $2.8\text{ s} = 2\pi\sqrt{\dfrac{1.5\text{ m}}{g}}$

# 15 Traveling Waves and Sound

## 15.1 The Wave Model

## 15.2 Traveling Waves

1. a. In your own words, define what a *transverse wave* is.

   b. Give an example of a wave that, from your own experience, you know is a transverse wave. What observations or evidence tells you this is a transverse wave?

2. a. In your own words, define what a *longitudinal wave* is.

   b. Give an example of a wave that, from your own experience, you know is a longitudinal wave. What observations or evidence tells you this is a longitudinal wave?

3. Three wave pulses travel along the same string. Rank in order, from largest to smallest, their wave speeds $v_1$, $v_2$, and $v_3$.

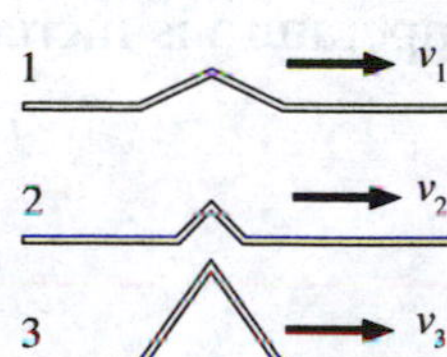


Order:

Explanation:

4. A wave pulse travels along a string at a speed of 200 cm/s. What will be the speed if:
**Note:** Each part below is independent and refers to changes made to the original string.
   a. The string's tension is doubled?

   b. The string's mass is quadrupled (but its length is unchanged)?

   c. The string's length is quadrupled (but its mass is unchanged)?

   d. The string's mass and length are both quadrupled?

5. Sound travels through a 300 K gas at 400 m/s. What will be the sound speed if the gas temperature is increased to 600 K? Explain.

## 15.3 Graphical and Mathematical Descriptions of Waves

6. Each figure below shows a snapshot graph at time $t = 0$ s of a wave pulse on a string. The pulse on the left is traveling to the right at 100 cm/s; the pulse on the right is traveling to the left at 100 cm/s. Draw snapshot graphs of the wave pulse at the times shown next to the axes.

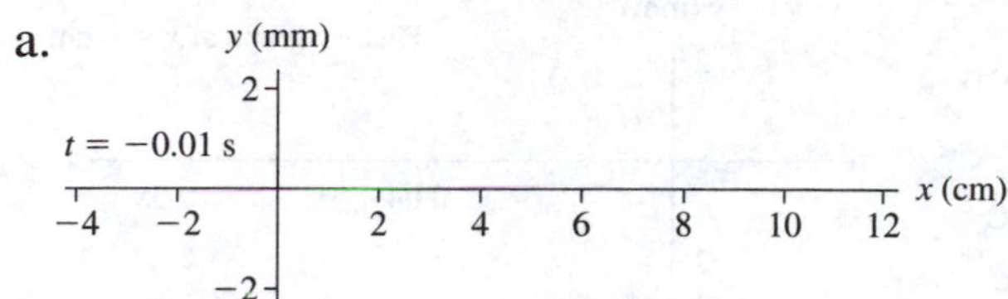


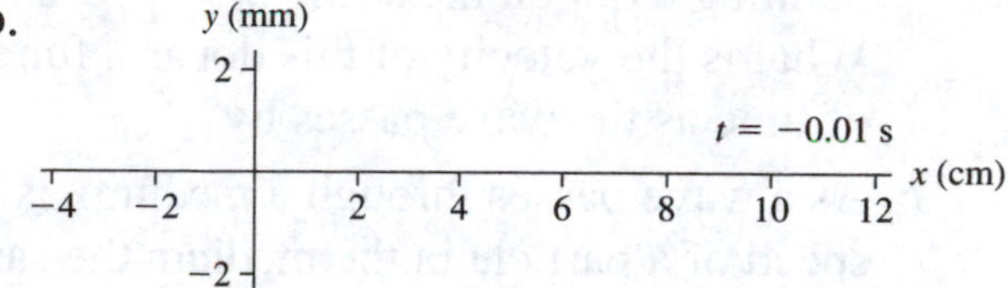


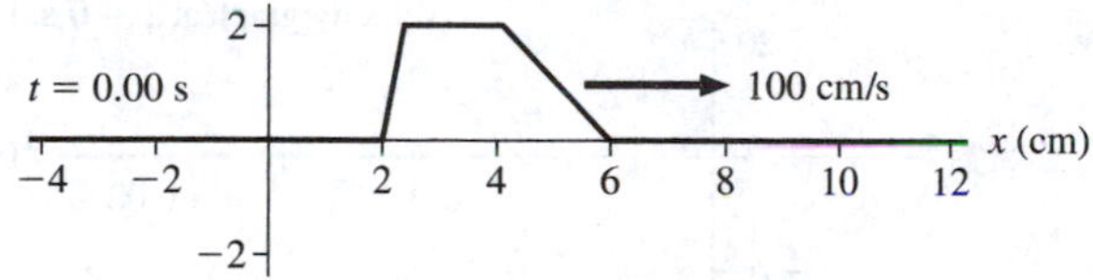


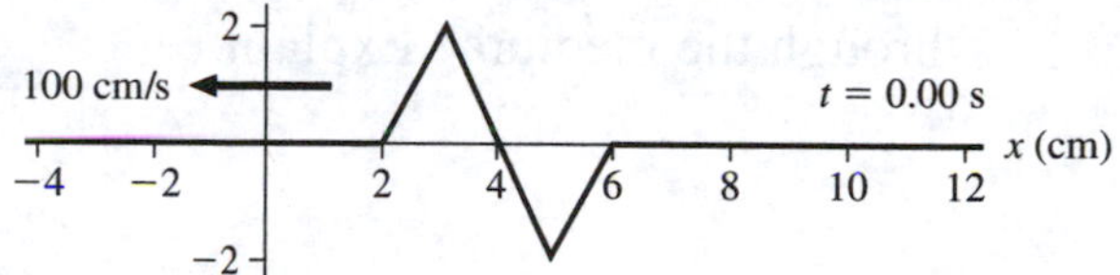


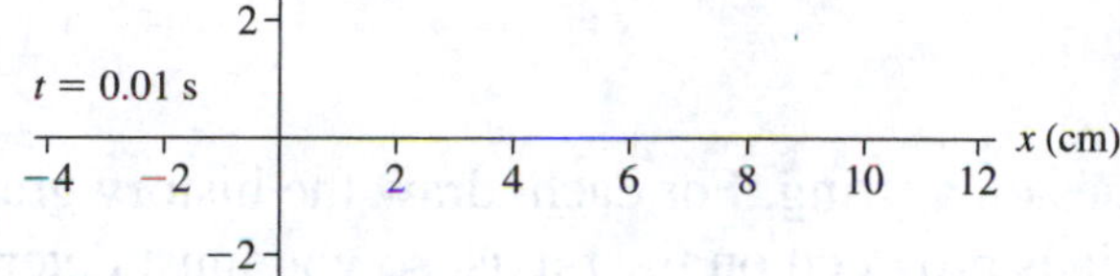


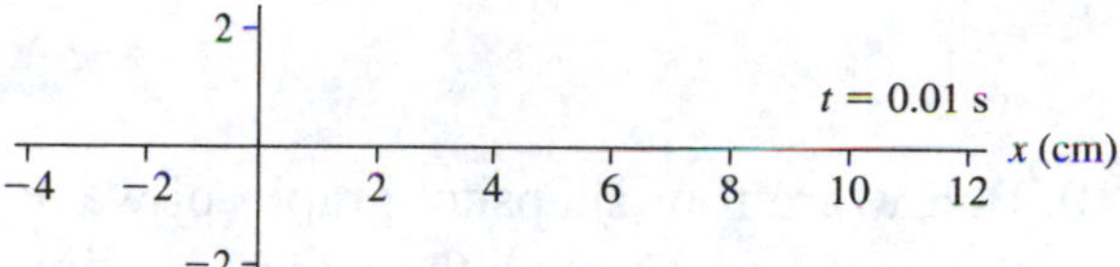


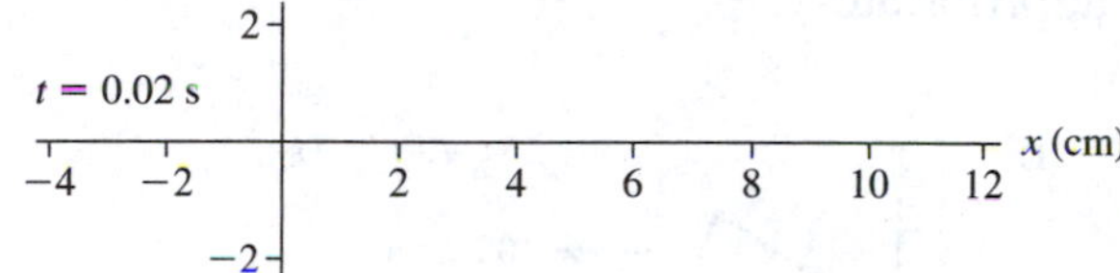


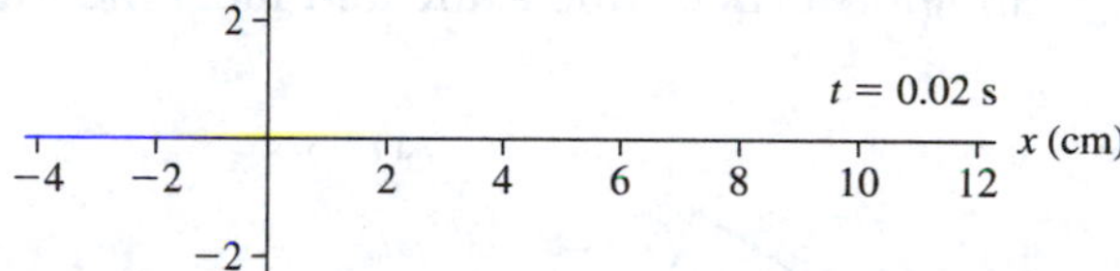


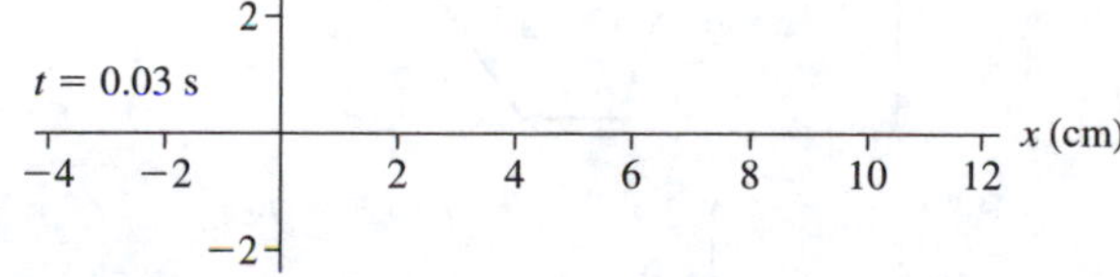


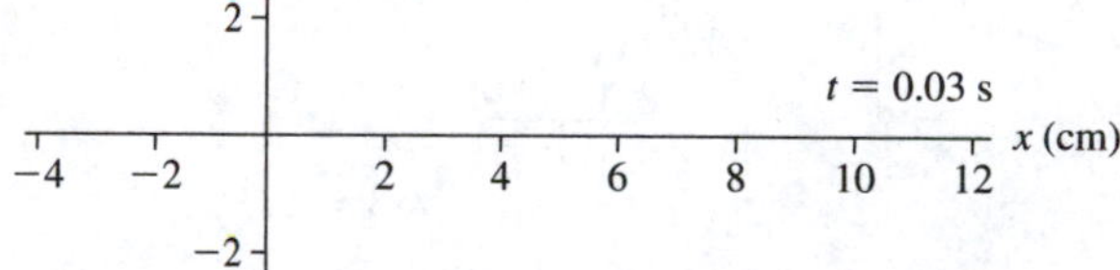


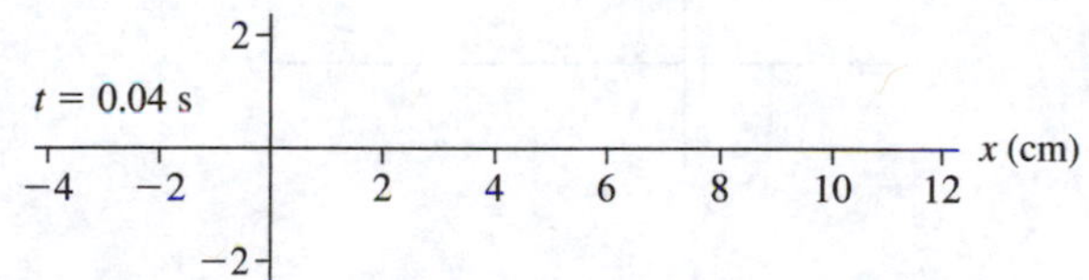


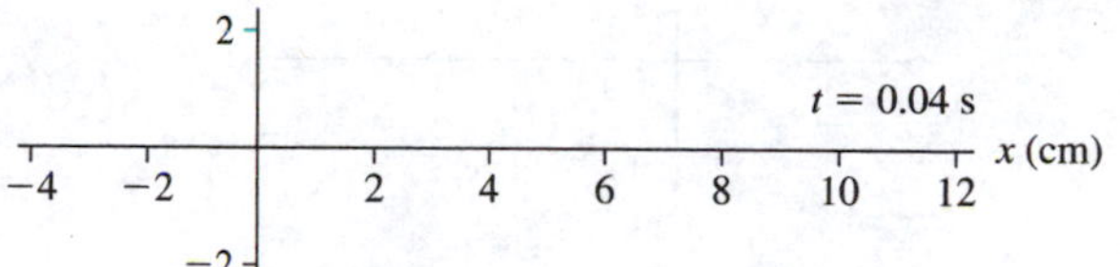


7. This snapshot graph is taken from Exercise 6a. On the axes below, draw the *history* graphs $y(x = 2\text{ cm}, t)$ and $y(x = 6\text{ cm}, t)$ showing the displacement at $x = 2$ cm and $x = 6$ cm as functions of time. Refer to your graphs in Exercise 6a to see what is happening at different instants of time.

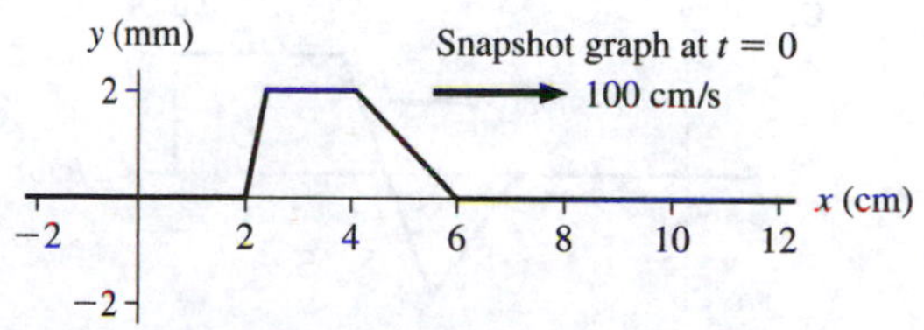


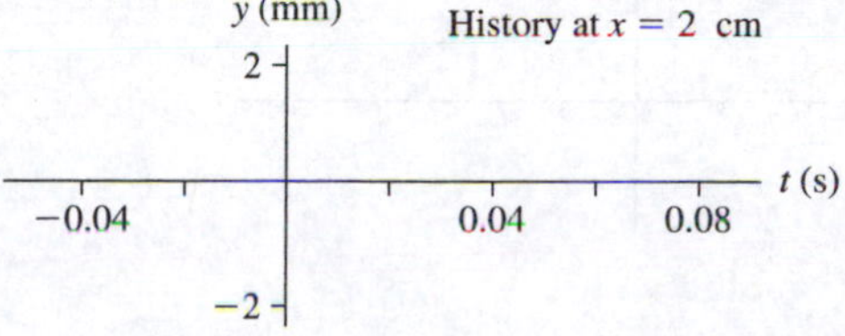


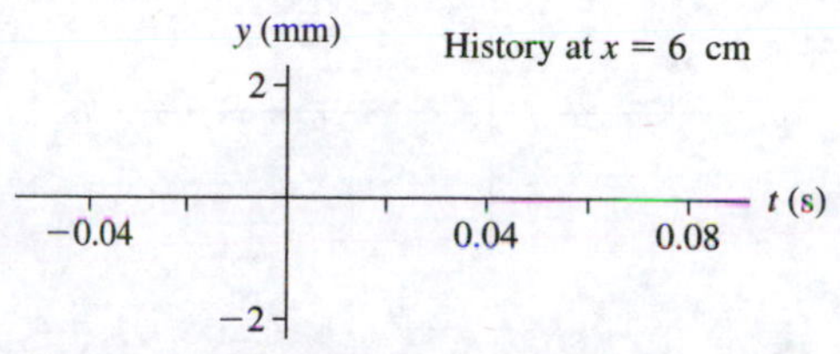

8. This snapshot graph is from Exercise 6b.
   a. Draw the history graph $y(x = 0 \text{ cm}, t)$ for this wave at the point $x = 0$ cm.
   b. Draw the *velocity*-versus-time graph for the piece of the string at $x = 0$ cm. Imagine painting a dot on the string at $x = 0$ cm. What is the velocity of this dot as a function of time as the wave passes by?
   c. As a wave passes through a medium, is the speed of a particle in the medium the same as or different from the speed of the wave through the medium? Explain.

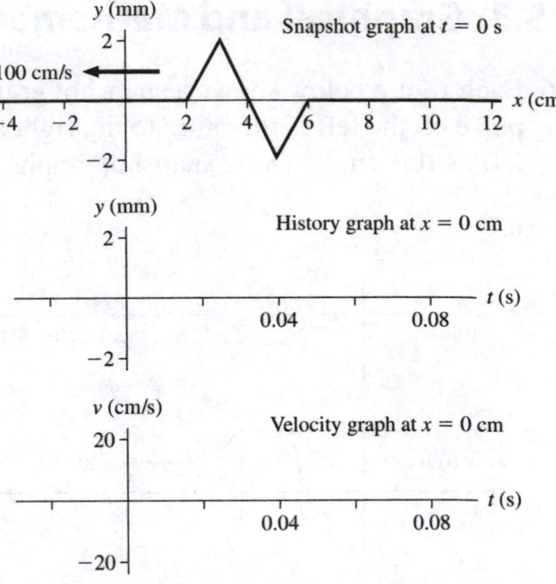


9. Below are four snapshot graphs of wave pulses on a string. For each, draw the history graph at the specified point on the x-axis. No time scale is provided on the t-axis, so you must determine an appropriate time scale and label the t-axis appropriately.

a. y At t = 0 s

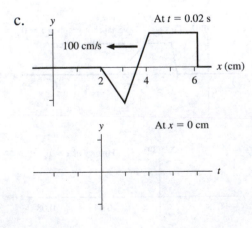

100 cm/s

x (cm) 2 4 6

y At x = 2 cm

t

b.

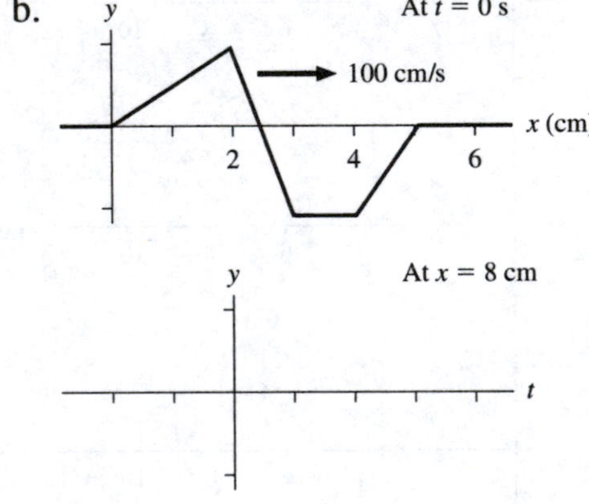


c. y At t = 0.02 s

100 cm/s

x (cm) 2 4 6

y At x = 0 cm

t

d.

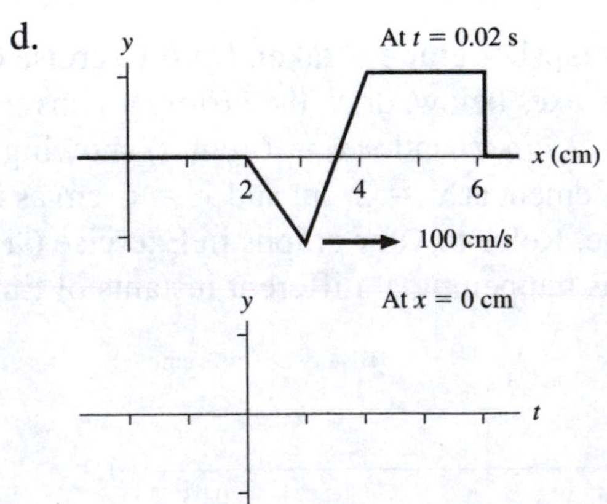

10. A history graph $y(x = 0 \text{ cm}, t)$ is shown for the $x = 0$ cm point on a string. The pulse is moving to the right at 100 cm/s.

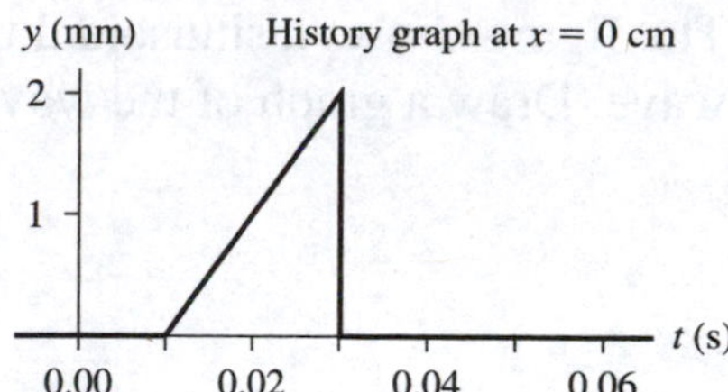


a. Does the $x = 0$ cm point on the string rise quickly and then fall slowly, or rise slowly and then fall quickly? Explain.

b. At what time does the leading edge of the wave pulse arrive at $x = 0$ cm?

c. At $t = 0$ s, how far is the leading edge of the wave pulse from $x = 0$ cm? Explain.

d. At $t = 0$ s, is the leading edge to the right or to the left of $x = 0$ cm?

e. At what time does the trailing edge of the wave pulse leave $x = 0$ cm?

f. At $t = 0$ s, how far is the trailing edge of the pulse from $x = 0$ cm? Explain.

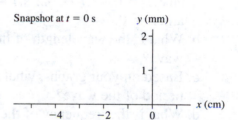

g. By referring to the answers you've just given, draw a snapshot graph $y(x = 0 \text{ s})$ showing the wave pulse on the string at $t = 0$ s.

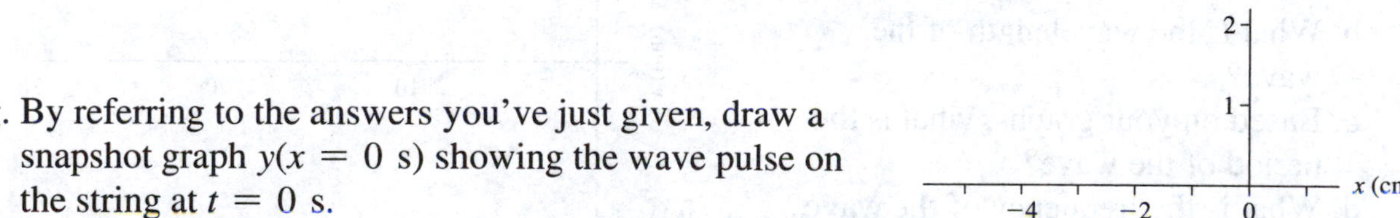


11. These are a history graph *and* a snapshot graph for a wave pulse on a string. They describe the same wave from two perspectives.

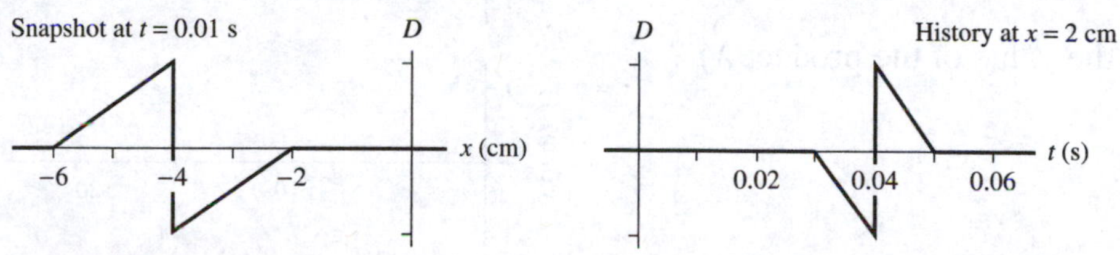


What is the speed of this wave?

12. The figure shows a sinusoidal traveling wave. Draw a graph of the wave if:

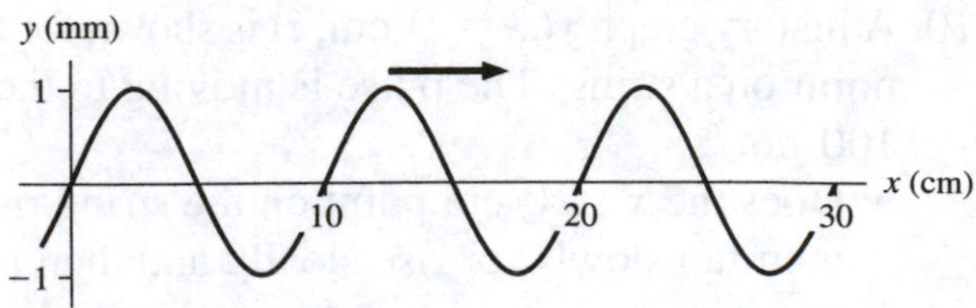


a. Its amplitude is halved and its wavelength is doubled.

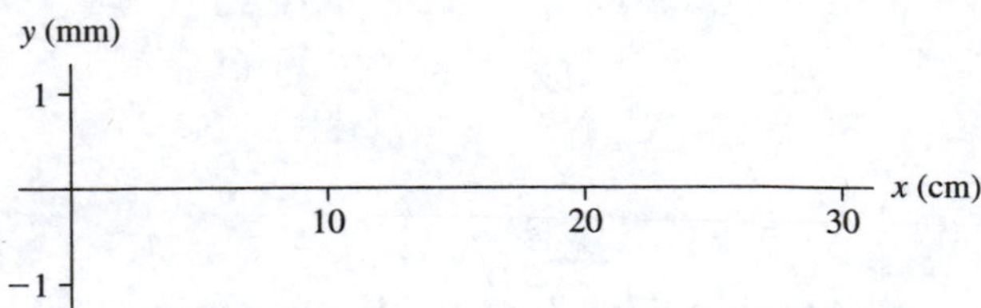


b. Its speed is doubled and its frequency is quadrupled.

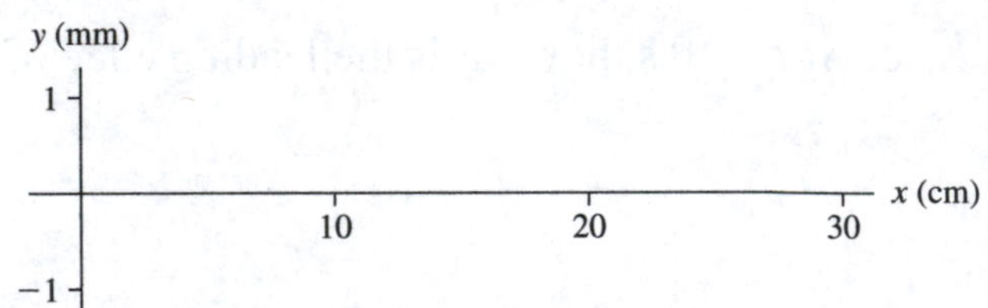


13. The wave shown at time $t = 0$ s is traveling to the right at a speed of 25 cm/s.

a. Draw snapshot graphs of this wave at times $t = 0.1$ s, $t = 0.2$ s, $t = 0.3$ s, and $t = 0.4$ s.

b. What is the wavelength of the wave? __________

c. Based on your graphs, what is the period of the wave? __________

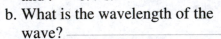

d. What is the frequency of the wave?

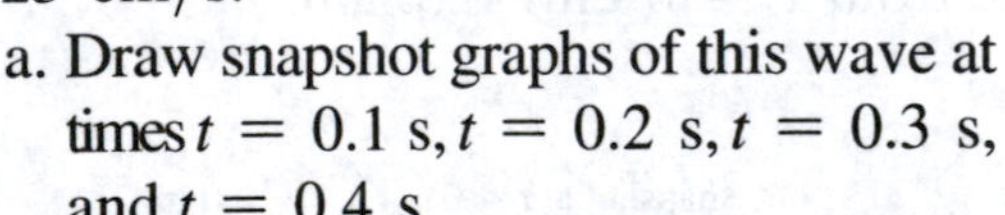

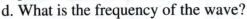

e. What is the value of the product $\lambda f$?

f. How does this value of $\lambda f$ compare to the speed of the wave?

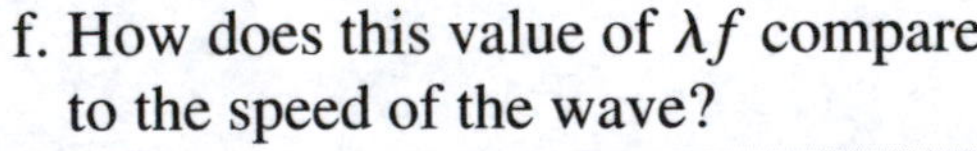

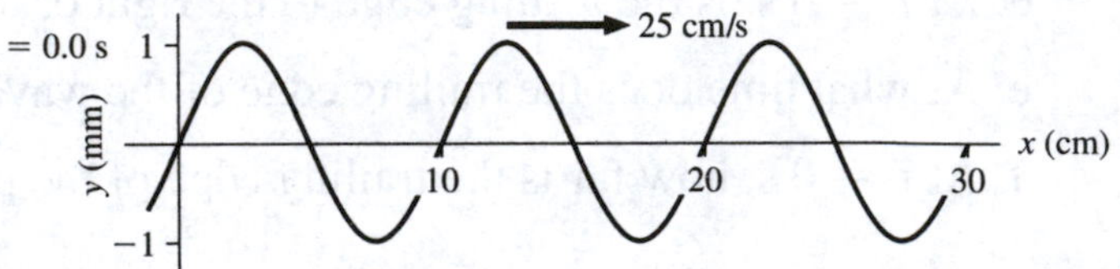


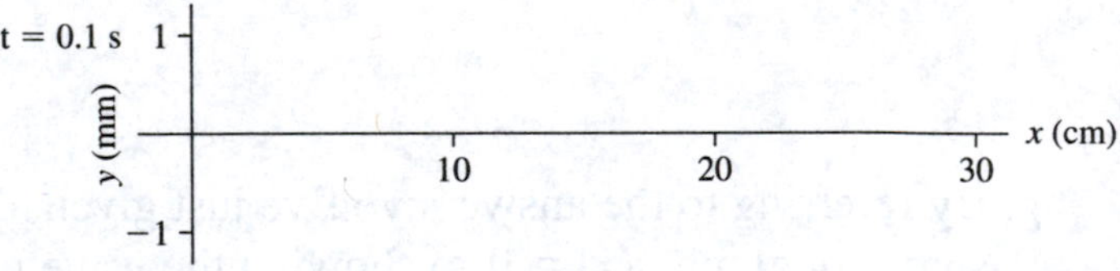


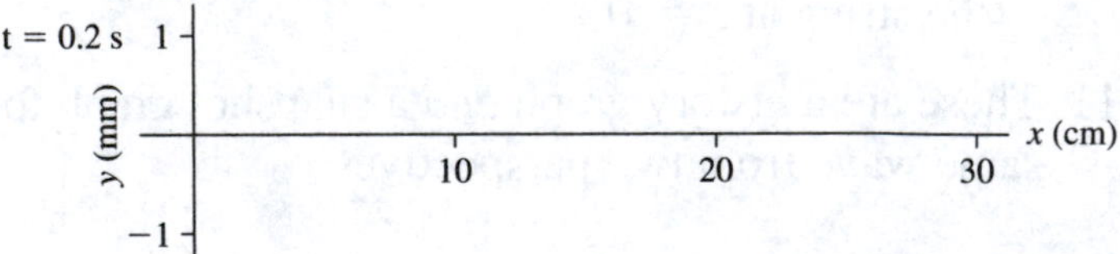


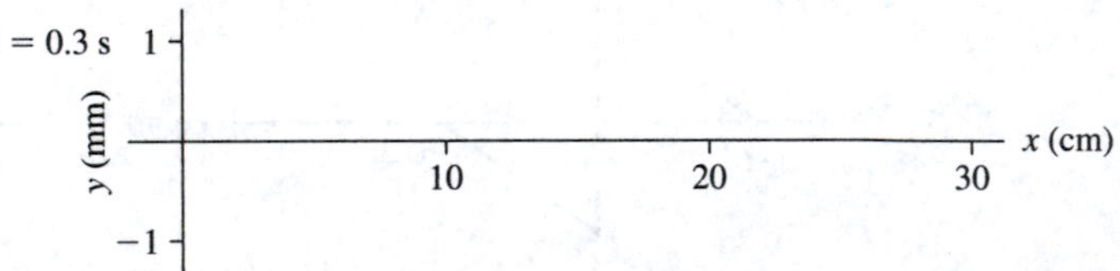


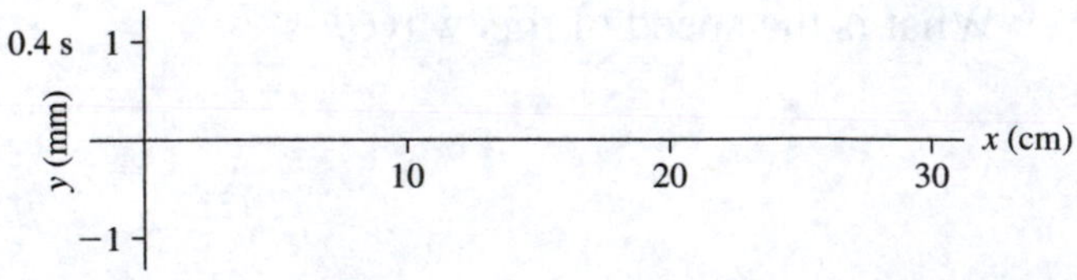

## 15.4 Sound and Light Waves

14. A horizontal Slinky is at rest on a table. A wave pulse is sent along the Slinky, causing the top of link 5 to move *horizontally* with the *displacement* from equilibrium shown in the graph.

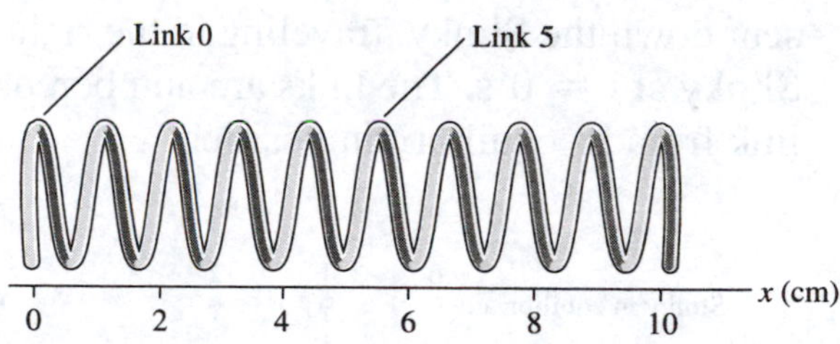


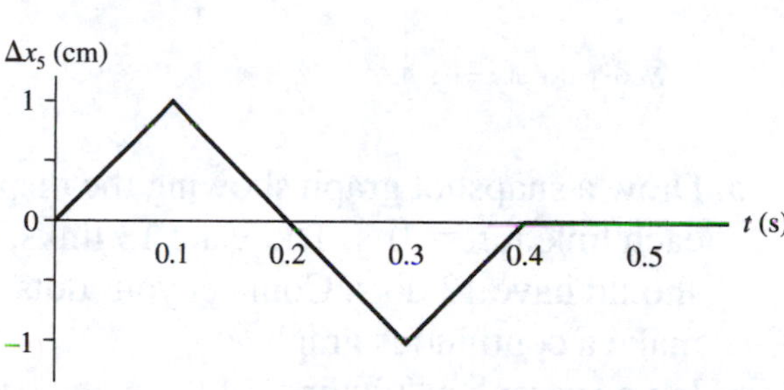


a. Is this a transverse or a longitudinal wave? Explain.

b. What is the *position* of link 5 at $t = 0.1$ s? ______

What is the *position* of link 5 at $t = 0.2$ s? ______

What is the *position* of link 5 at $t = 0.3$ s? ______

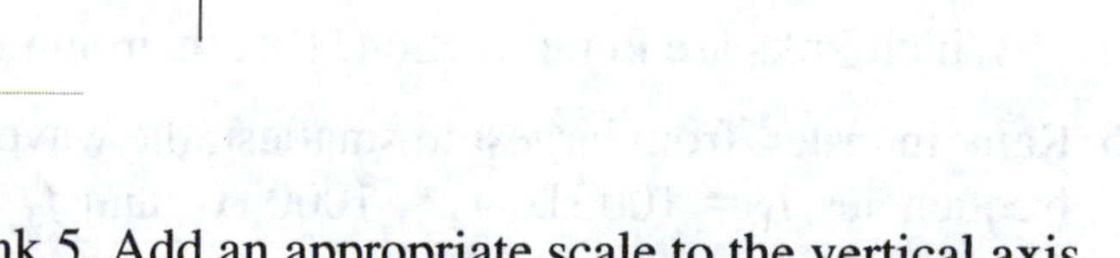

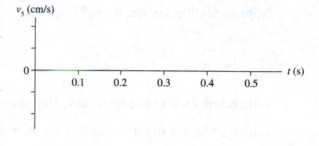

**Note:** *Position*, not displacement.

c. Draw a velocity-versus-time graph of link 5. Add an appropriate scale to the vertical axis. (Recall how velocity graphs are related to the slopes of position graphs.)

d. Can you determine, from the information given, the speed of the wave? If so, give the speed and explain how you found it. If not, why not?

15. We can use a series of dots to represent the positions of the links in a Slinky. The top set of dots shows a Slinky in equilibrium with a 1-cm spacing between the links. A wave pulse is sent down the Slinky, traveling to the right at 10 cm/s. The second set of dots shows the Slinky at $t = 0$ s. The links are numbered, and you can measure the displacement $\Delta x$ of each link from its equilibrium position.

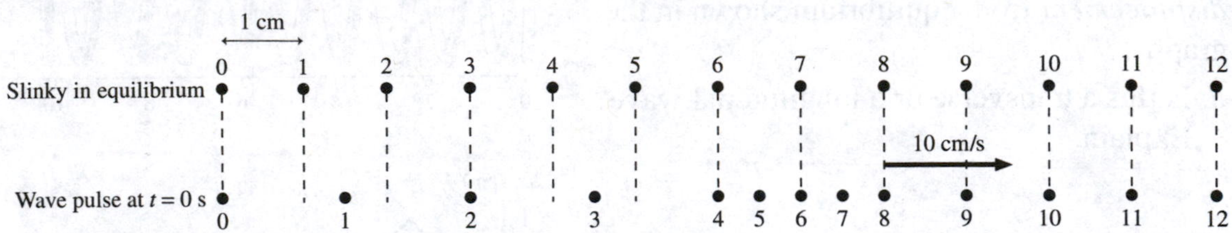


a. Draw a snapshot graph showing the displacement $\Delta x$ of each link at $t = 0$ s. There are 13 links, so your graph should have 13 dots. Connect your dots with lines to make a continuous graph.

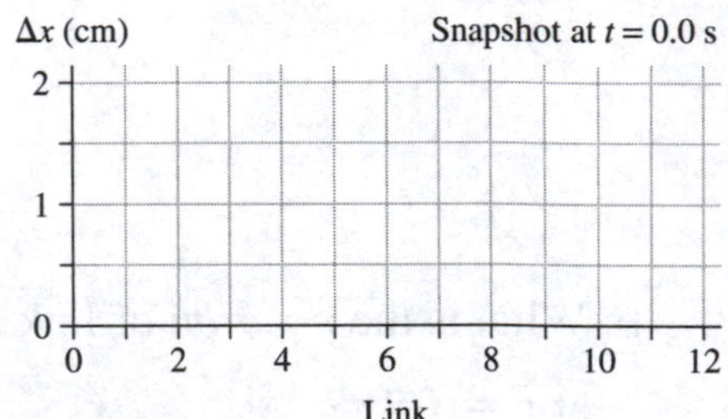


b. Is your graph a "picture" of the wave or a "representation" of the wave? Explain.

c. Which links are in compression? (list their numbers)

Which links are in rarefaction? (list their numbers)

16. Rank in order, from largest to smallest, the wavelengths $\lambda_1$ to $\lambda_3$ for sound waves having frequencies $f_1 = 100$ Hz, $f_2 = 1000$ Hz, and $f_3 = 10{,}000$ Hz.

Order:

Explanation:

## 15.5 Energy and Intensity

## 15.6 Loudness of Sound

17. The figure shows the path of a light wave past a series of equally spaced *transparent* grids. The portion of the first two grids that would be illuminated by the light is represented by the shaded area in the first two grids.
    a. Complete the figure by shading in the spaces that would be illuminated in the remaining two grids.

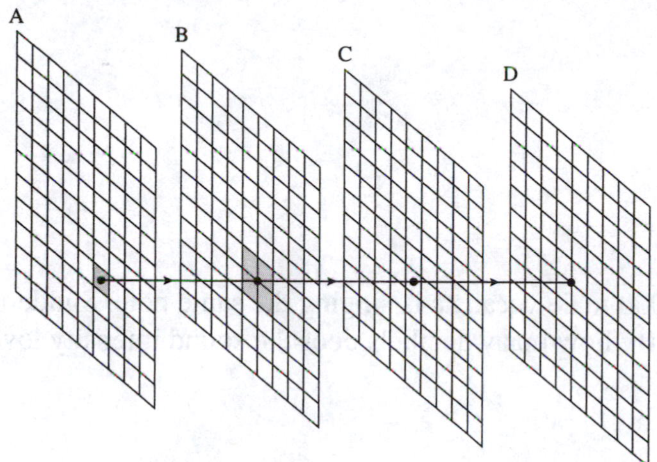


    b. Is the energy of the light wave passing through the fourth transparent grid (D) greater than, less than, or equal to the energy passing through the first transparent grid (A)? Explain.

    c. What is the ratio $I_D/I_A$ of the intensity of the light at the fourth grid (D) to the intensity at the first grid (A)? Explain.

18. A laser beam has intensity $I_0$.
    a. What is the intensity, in terms of $I_0$, if a lens focuses the laser beam to $\frac{1}{10}$ its initial diameter?

    b. What is the intensity, in terms of $I_0$, if a lens defocuses the laser beam to 10 times its initial diameter?

19. Sound wave A delivers 2 J of energy in 2 s. Sound wave B delivers 10 J of energy in 5 s. Sound wave C delivers 2 mJ of energy in 1 ms. Rank in order, from largest to smallest, the sound powers $P_A$, $P_B$, and $P_C$ of these three sound waves.

Order:

Explanation:

20. A giant chorus of 1000 male vocalists is singing the same note. Suddenly, 999 vocalists stop, leaving one soloist. By how many decibels does the sound intensity level decrease? Explain.

## 15.7 The Doppler Effect and Shock Waves

21. Five expanding wave fronts from a moving sound source are shown. The dots represent the centers of the respective circular wave fronts, which is the location of the source when that wave front was emitted. The frequency of the sound emitted by the source is constant.

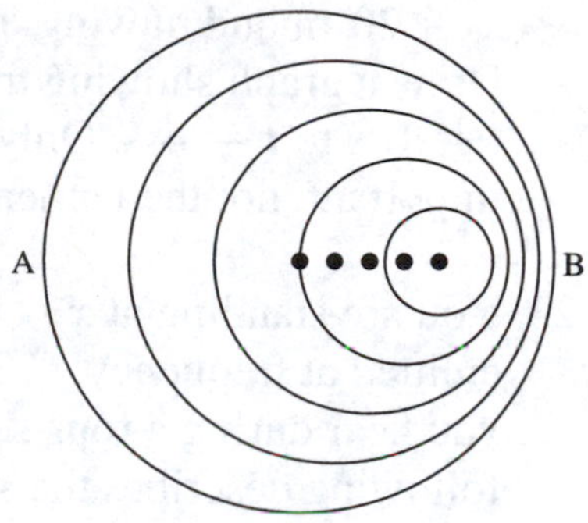


a. Indicate on the figure the direction of motion of the source. Which sound wave front was produced first? How do you know? Explain.

b. Do the observers at locations A and B hear the same frequency of sound? If not, which one hears a higher frequency? Explain.

c. Assume that the sound wave you identified in part a as the first wave front produced marks the beginning of the sound. Do the observers at A and B first hear the sound at the same time? If not, which one hears the sound first? Explain.

d. The speed of sound in the medium is $v$. Is the speed $v_s$ of the source greater than, less than, or equal to $v$? Explain.

22. You are standing at $x = 0$ m, listening to a sound that is emitted at a frequency $f_s$. At $t = 0$ s, the sound source is at $x = 20$ m and moving toward you at a steady 10 m/s. Draw a graph showing the frequency you hear from $t = 0$ s to $t = 4$ s. Only the shape of the graph is important, not the numerical values of $f$.

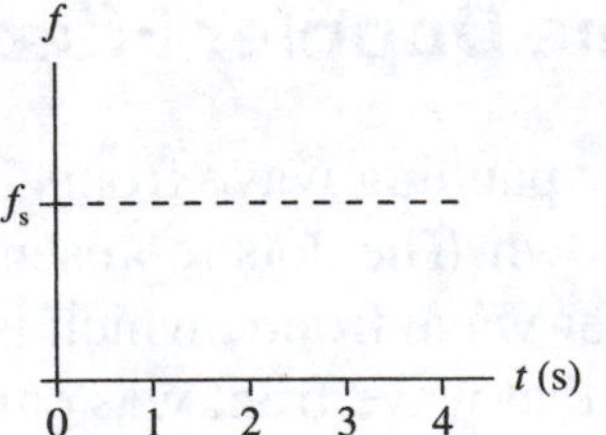

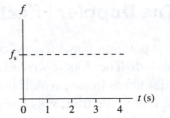

23. You are standing at $x = 0$ m, listening to a sound that is emitted at frequency $f_s$. The graph shows the frequency you hear during a four-second interval. Which of the following describes the sound source?

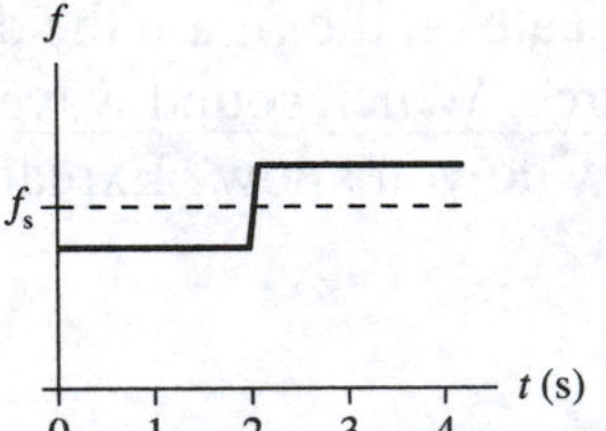


i. It moves from left to right and passes you at $t = 2$ s.
ii. It moves from right to left and passes you at $t = 2$ s.
iii. It moves toward you but doesn't reach you. It then reverses direction at $t = 2$ s.
iv. It moves away from you until $t = 2$ s. It then reverses direction and moves toward you but doesn't reach you.

Explain your choice.

24. You are standing at $x = 0$ m, listening to seven identical sound sources. At $t = 0$ s, all seven are at $x = 343$ m and moving as shown below. The sound from all seven will reach your ear at $t = 1$ s.

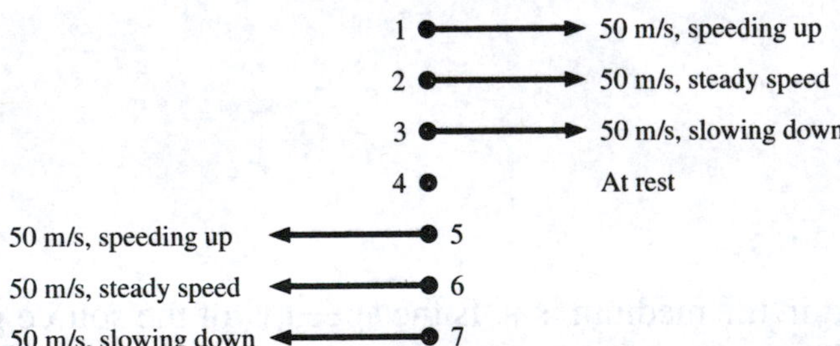


Rank in order, from highest to lowest, the seven frequencies $f_1$ to $f_7$ that you hear at $t = 1$ s.

Order:

Explanation:

# 16 Superposition and Standing Waves

## 16.1 The Principle of Superposition

1. Two pulses on a string are approaching each other at 10 m/s. Draw snapshot graphs of the string at the three times indicated.

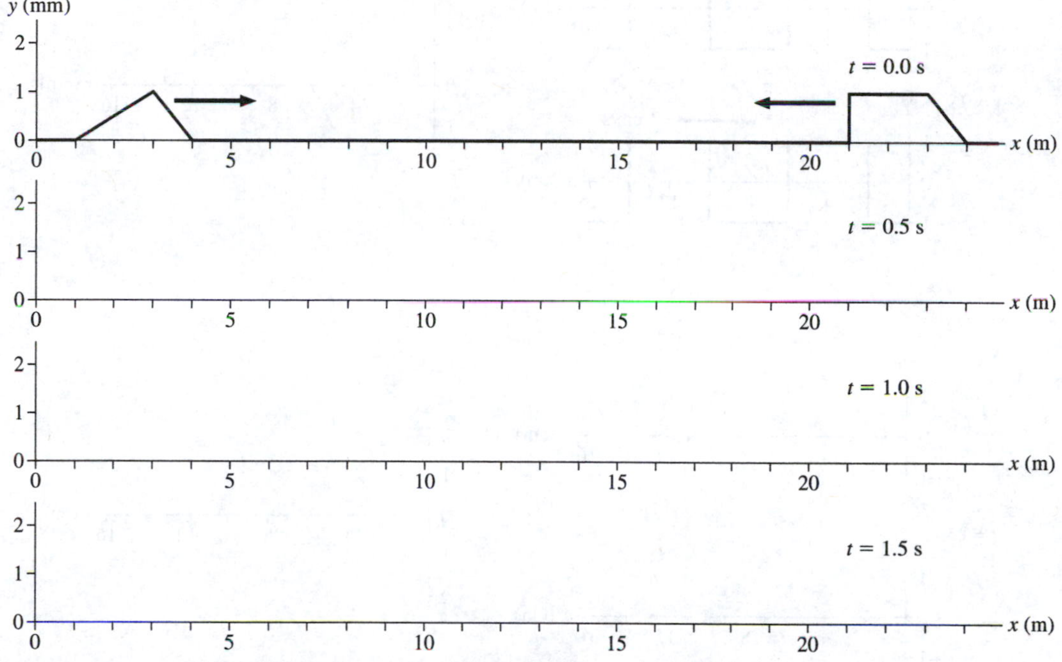


2. Two pulses on a string are approaching each other at 10 m/s. Draw a snapshot graph of the string at $t = 1$ s.

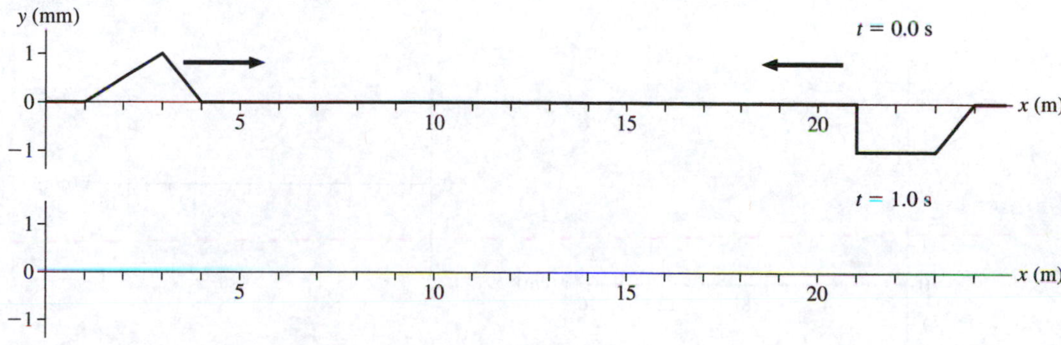

## 16.2 Standing Waves

3. Two waves are traveling in opposite directions along a string. Each has a speed of 1 cm/s, and an amplitude of 1 cm. The first set of graphs below shows each wave at $t = 0$ s.
   a. On the axes at the right, draw the superposition of these two waves at $t = 0$ s.
   b. On the axes at the left, draw each of the two displacements every 2 s until $t = 8$ s. The waves extend beyond the graph edges, so new pieces of the wave will move in.
   c. On the axes at the right, draw the superposition of the two waves at the same instant.

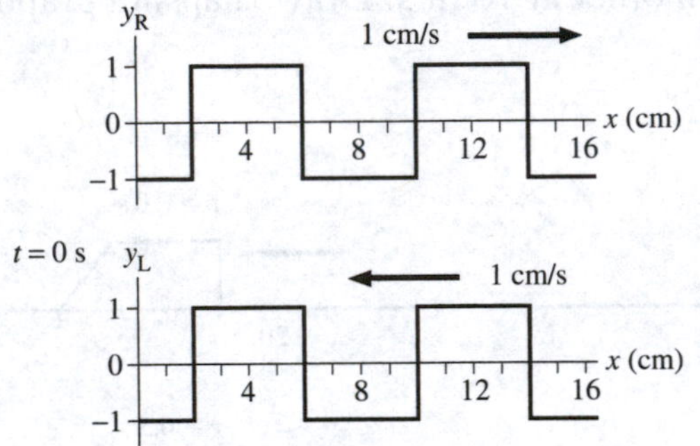


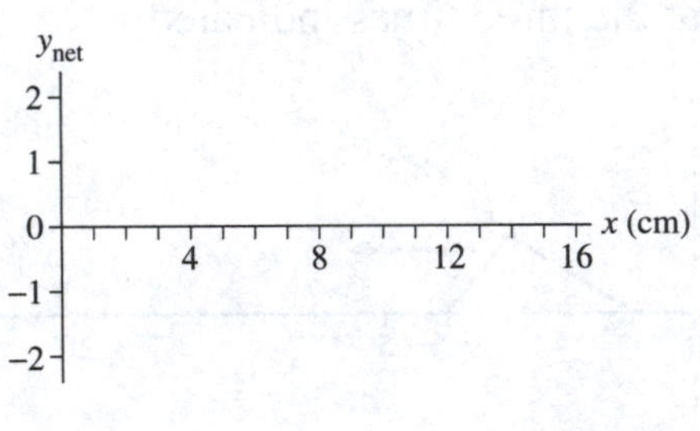


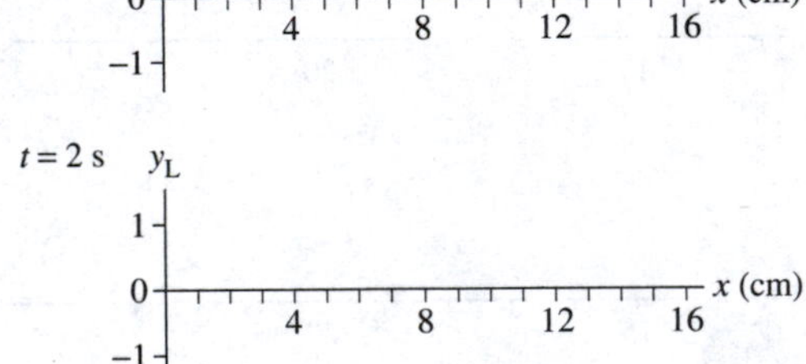


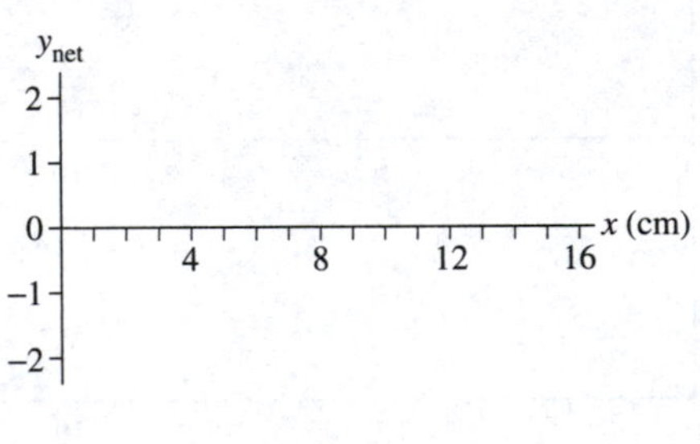


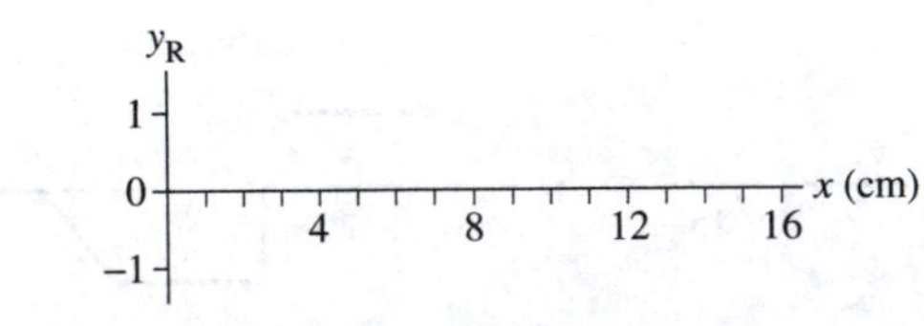


$t = 4$ s

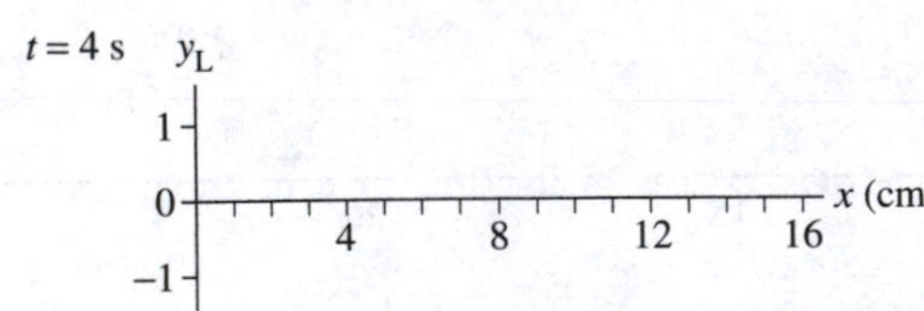


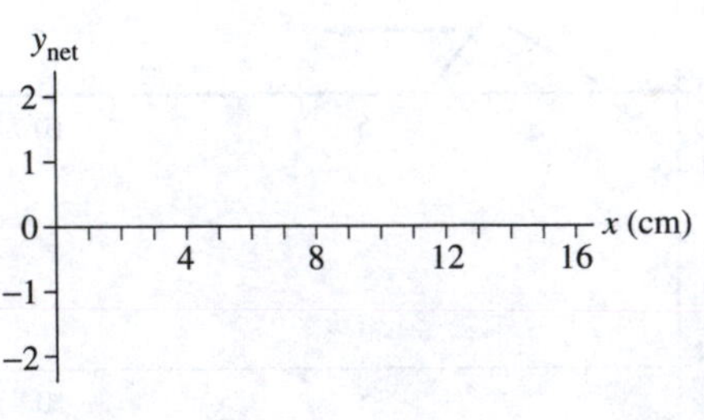


*(Continues next page)*

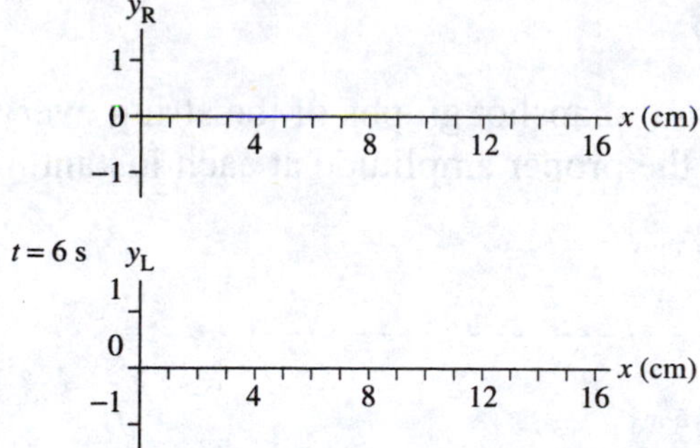
$y_R$
1
0
−1
x (cm)
4
8
12
16
t = 6 s
$y_L$
1
0
−1
x (cm)
4
8
12
16

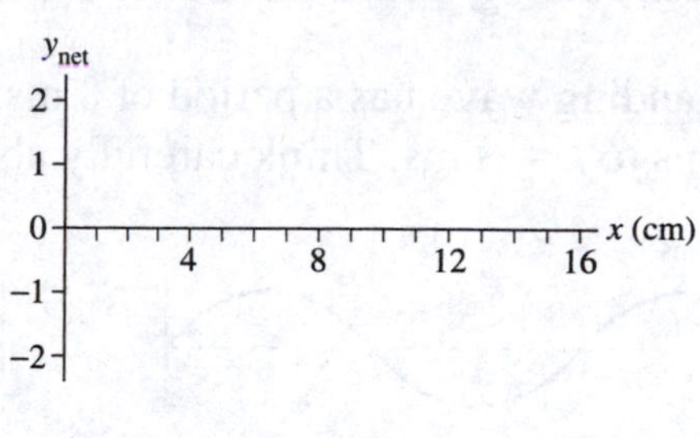
$y_{net}$
2
1
0
−1
−2
x (cm)
4
8
12
16

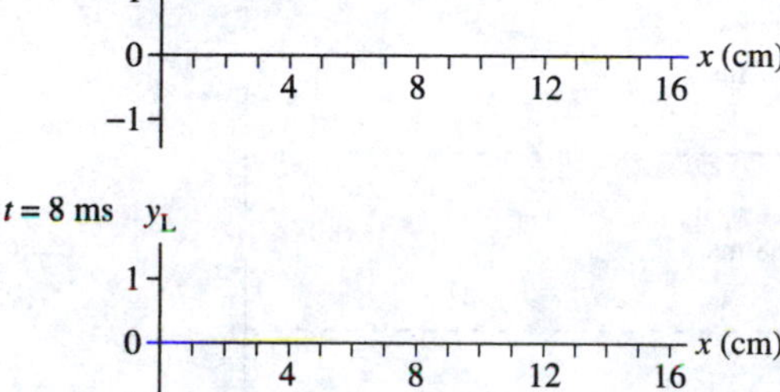
$y_R$
1
0
−1
x (cm)
4
8
12
16
t = 8 ms
$y_L$
1
0
−1
x (cm)
4
8
12
16

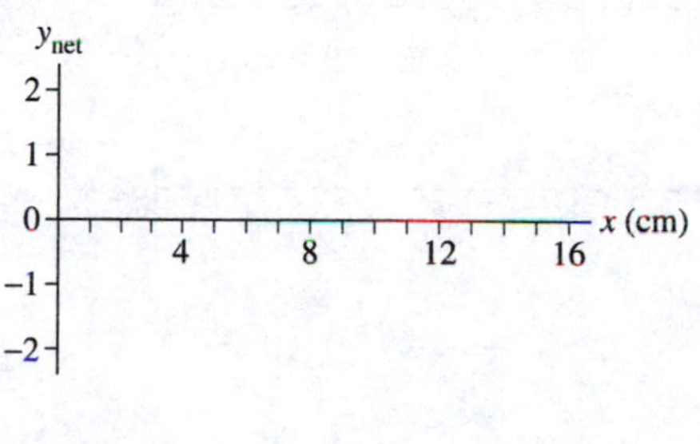
$y_{net}$
2
1
0
−1
−2
x (cm)
4
8
12
16

## 16.3 Standing Waves on a String

4. This standing wave has a period of 8 ms. Draw snapshot graphs of the string every 1 ms from $t = 1$ ms to $t = 8$ ms. Think carefully about the proper amplitude at each instant.

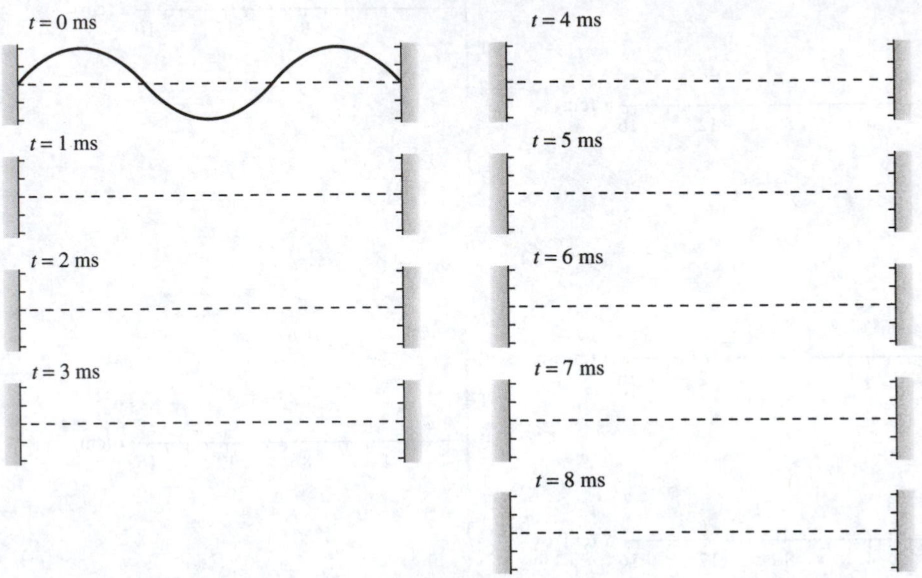


5. The figure shows a standing wave on a string. It has frequency $f$.

a. Draw the standing wave if the frequency is changed to $\frac{2}{3}f$ and to $\frac{3}{2}f$.

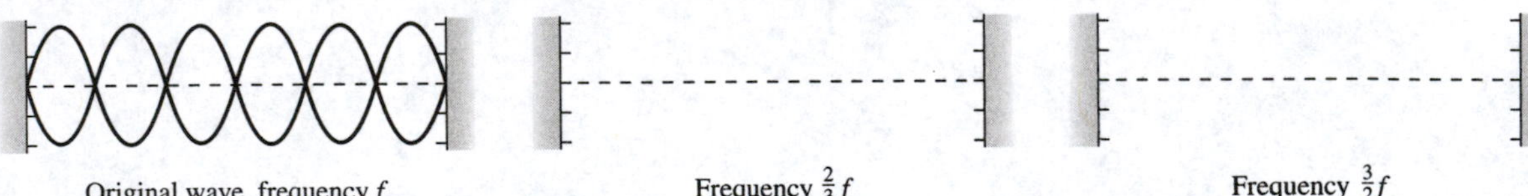

Original wave, frequency $f$ Frequency $\frac{2}{3}f$ Frequency $\frac{3}{2}f$

b. Is there a standing wave if the frequency is changed to $\frac{1}{4}f$? If so, how many antinodes does it have? If not, why not?

6. The figure shows a standing wave on a string.

a. Draw the standing wave if the tension is quadrupled while the frequency is held constant.

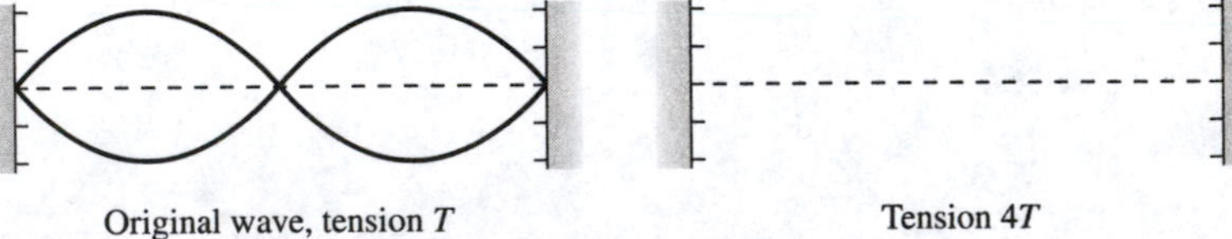

Original wave, tension $T$ Tension $4T$

b. Suppose the tension is merely doubled while the frequency remains constant. Will there be a standing wave? If so, how many antinodes will it have? If not, why not?

## 16.4 Standing Sound Waves

7. The picture shows a *displacement* graph (not a pressure graph) of a standing sound wave in a 32-mm-long tube of air that is open at both ends. That is, the graph shows how far a molecule is displaced from its equilibrium position.

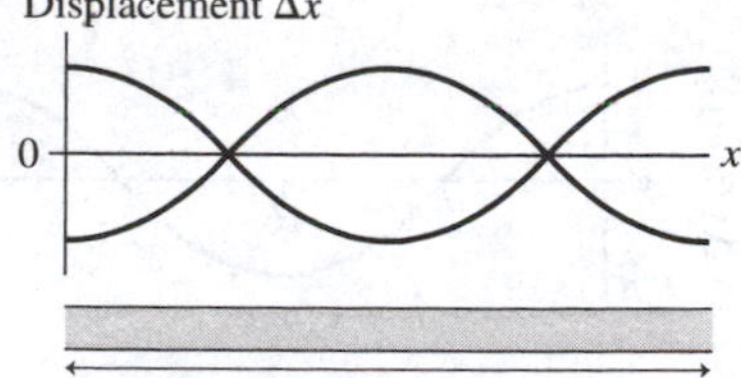


   a. Which mode (value of $m$) standing wave is this?
   b. Are the air molecules vibrating vertically or horizontally? Explain.

   c. At what distances from the left end of the tube do the molecules oscillate with maximum amplitude?

8. The purpose of this exercise is to visualize the motion of the air molecules for the standing wave of Exercise 7. On the next page are nine graphs, every one-eighth of a period from $t = 0$ to $t = T$. Each graph represents the displacements at that instant of time of the molecules in a 32-mm-long tube. Positive values are displacements to the right, negative values are displacements to the left.
   a. Consider nine air molecules that, in equilibrium, are 4 mm apart and lie along the axis of the tube. The top picture on the right shows these molecules in their equilibrium positions. The dotted lines down the page—spaced 4 mm apart—are reference lines showing the equilibrium positions. Read each graph carefully, then draw nine dots to show the positions of the nine air molecules at each instant of time. The first one, for $t = 0$, has already been done to illustrate the procedure.

      **Note:** It's a good approximation to assume that the left dot moves in the pattern 4, 3, 0, −3, −4, −3, 0, 3, 4 mm; the second dot in the pattern 3, 2, 0, −2, −3, −2, 0, 2, 3 mm; and so on.
   b. At what times does the air reach maximum compression, and where does it occur?

      Max compression at time ________ Max compression at position ________

   c. What is the relationship between the positions of maximum compression and the nodes of the standing wave?

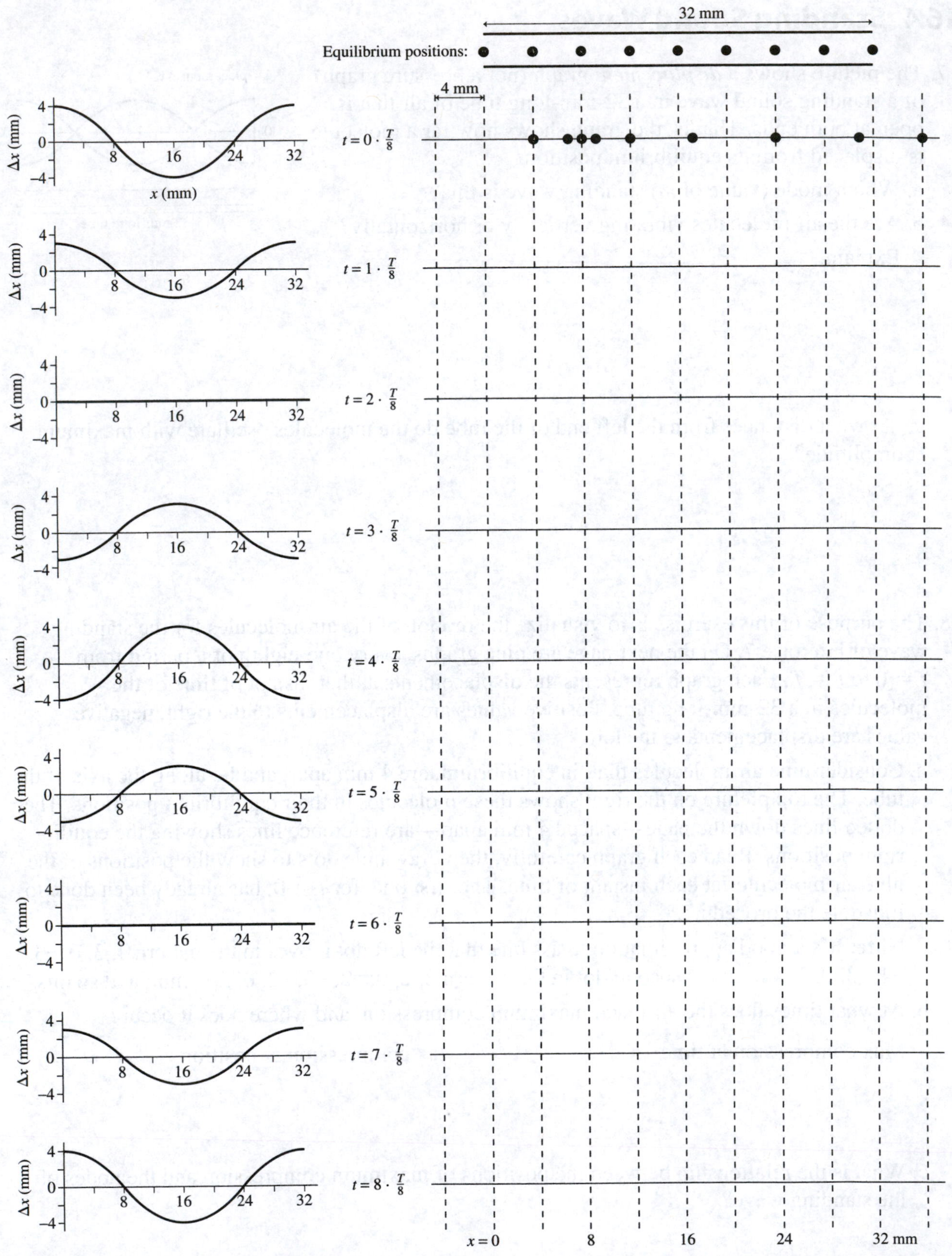
32 mm
Equilibrium positions:
4 mm
Δx (mm)
x (mm)
4
0
−4
8
16
24
32
t = 0 · T/8
t = 1 · T/8
t = 2 · T/8
t = 3 · T/8
t = 4 · T/8
t = 5 · T/8
t = 6 · T/8
t = 7 · T/8
t = 8 · T/8
x = 0
8
16
24
32 mm

## 16.5 Speech and Hearing

## 16.6 The Interference of Waves from Two Sources

9. The figure shows a snapshot graph at $t = 0$ s of loudspeakers emitting triangular-shaped sound waves. Speaker 2 can be moved forward or backward along the axis. Both speakers vibrate in phase at the same frequency. The second speaker is drawn below the first, so that the figure is clear, but you want to think of the two waves as overlapped as they travel along the *x*-axis.
   a. On the left set of axes, draw the $t = 0$ s snapshot graph of the second wave if speaker 2 is placed at each of the positions shown. The first graph, with $x_{speaker} = 2$ m, is already drawn.

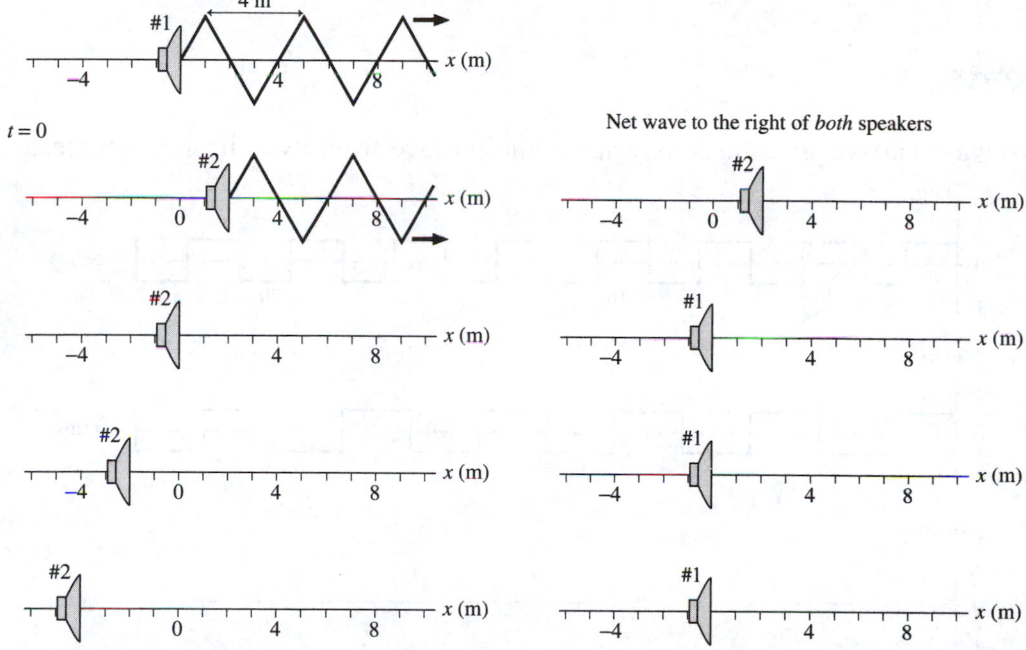


   b. On the right set of axes, draw the superposition $\Delta p = \Delta p_1 + \Delta p_2$ of the waves from the two speakers. $\Delta p$ exists only to the right of *both* speakers. It is the net wave traveling to the right.
   c. What distance or distances $\Delta d$ between the speakers give constructive interference?

   d. What are the values of $\Delta d/\lambda$, the ratio of path-length distance to wavelength, at the points of constructive interference?

   e. What distance or distances $\Delta d$ between the speakers give destructive interference?

   f. What are the values of $\Delta d/\lambda$ at the points of destructive interference?

10. The figure shows the wave-front pattern emitted by two loudspeakers.

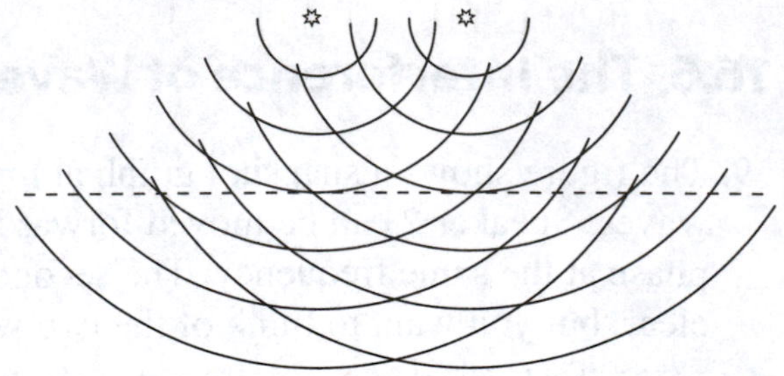

a. Draw a dot • at points where there is constructive interference. These are points where two crests overlap *or* two troughs overlap.

b. Draw an open circle ○ at points where there is destructive interference. These are points where a crest overlaps a trough.

## 16.7 Beats

11. The two waves arrive simultaneously at a point in space from two different sources.

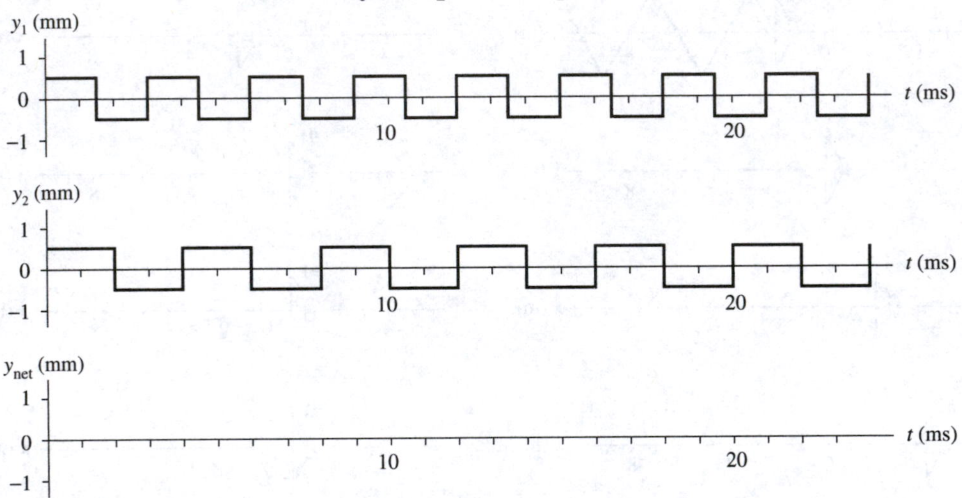


a. Period of wave 1? ________ Frequency of wave 1? ________

b. Period of wave 2? ________ Frequency of wave 2? ________

c. Draw the graph of the net wave at this point on the third set of axes. Be accurate, use a ruler!

d. Period of the net wave? ________ Frequency of the net wave? ________

e. Is the frequency of the superposition what you would expect as a beat frequency? Explain.

## DYNAMICS WORKSHEET Name ______________________ Problem ________

### PREPARE

- List knowns. Identify what you're trying to find.
- Identify forces and draw a free-body diagram.
- Draw a pictorial representation for problems with motion: Show important points in the motion, establish a coordinate system, define symbols, draw a motion diagram.

#### Known

#### Find

### SOLVE

Start with Newton's first or second law in component form, adding other information as needed to solve the problem.

### ASSESS

Have you answered the question?
Do you have correct units, signs, and significant figures?
Is your answer reasonable?

## DYNAMICS WORKSHEET Name ______________________________ Problem ______

### PREPARE

- List knowns. Identify what you're trying to find.
- Identify forces and draw a free-body diagram.
- Draw a pictorial representation for problems with motion: Show important points in the motion, establish a coordinate system, define symbols, draw a motion diagram.

**Known**

**Find**

### SOLVE

Start with Newton's first or second law in component form, adding other information as needed to solve the problem.

### ASSESS

Have you answered the question?
Do you have correct units, signs, and significant figures?
Is your answer reasonable?

## DYNAMICS WORKSHEET Name ______________________ Problem ________

### PREPARE

- List knowns. Identify what you're trying to find.
- Identify forces and draw a free-body diagram.
- Draw a pictorial representation for problems with motion: Show important points in the motion, establish a coordinate system, define symbols, draw a motion diagram.

**Known**

**Find**

### SOLVE

Start with Newton's first or second law in component form, adding other information as needed to solve the problem.

### ASSESS

Have you answered the question?
Do you have correct units, signs, and significant figures?
Is your answer reasonable?

## DYNAMICS WORKSHEET Name ______________________ Problem ______

### PREPARE

- List knowns. Identify what you're trying to find.
- Identify forces and draw a free-body diagram.
- Draw a pictorial representation for problems with motion: Show important points in the motion, establish a coordinate system, define symbols, draw a motion diagram.

**Known**

**Find**

### SOLVE

Start with Newton's first or second law in component form, adding other information as needed to solve the problem.

### ASSESS

Have you answered the question?
Do you have correct units, signs, and significant figures?
Is your answer reasonable?

## MOMENTUM WORKSHEET Name ______________________ Problem ________

### PREPARE

- Identify the system and any external forces.
- Sketch "before and after."
- Define coordinates.
- List knowns. Identify what you're trying to find.

**Known**

**Find**

- Is momentum conserved? ________

### SOLVE

Start with conservation of momentum or the impulse-momentum theorem, using Newton's laws or kinematics as needed.

### ASSESS

Have you answered the question?
Do you have correct units, signs, and significant figures?
Is your answer reasonable?

## MOMENTUM WORKSHEET Name ______________________ Problem ________

### PREPARE

- Identify the system and any external forces.
- Sketch "before and after."
- Define coordinates.
- List knowns. Identify what you're trying to find.

**Known**

**Find**

- Is momentum conserved? __________

### SOLVE

Start with conservation of momentum or the impulse-momentum theorem, using Newton's laws or kinematics as needed.

### ASSESS

Have you answered the question?

Do you have correct units, signs, and significant figures?

Is your answer reasonable?

## ENERGY WORKSHEET

Name ____________________ Problem ______

### PREPARE

- Identify the system.
- Sketch "before and after."
- Define coordinates.
- List knowns. Identify what you're trying to find.

**Known**

**Find**

- Is the system isolated? ____________ Which energies change? ____________________

### SOLVE

Start with conservation of energy, adding other information and techniques as needed to solve the problem.

### ASSESS

Have you answered the question?
Do you have correct units, signs, and significant figures?
Is your answer reasonable?

## ENERGY WORKSHEET

Name ______________________ Problem ________

PREPARE

- Identify the system.
- Sketch "before and after."
- Define coordinates.
- List knowns. Identify what you're trying to find.

**Known**

**Find**

- Is the system isolated? __________ Which energies change? __________

SOLVE

Start with conservation of energy, adding other information and techniques as needed to solve the problem.

ASSESS

Have you answered the question?
Do you have correct units, signs, and significant figures?
Is your answer reasonable?